Ecological Entanglements in the Anthropocene

Ecocritical Theory and Practice

Series Editor: Douglas A. Vakoch, California Institute of Integral Studies, USA

Ecocritical Theory and Practice highlights innovative scholarship at the interface of literary/cultural studies and the environment, seeking to foster an ongoing dialogue between academics and environmental activists.

Recent Titles

Ecological Entanglements in the Anthropocene, edited by Nicholas Holm and Sy Taffel
Ecocriticism, Ecology, and the Cultures of Antiquity, edited by Christopher Schliephake
Ecotheology and Nonhuman Ethics in Society: A Community of Compassion, edited by Melissa Brotton
The Ethics and Rhetoric of Invasion Ecology, edited by James Stanescu and Kevin Cummings

Ecological Entanglements in the Anthropocene

Nicholas Holm and Sy Taffel

LEXINGTON BOOKS
Lanham • Boulder • New York • London

Published by Lexington Books
An imprint of The Rowman & Littlefield Publishing Group, Inc.
4501 Forbes Boulevard, Suite 200, Lanham, Maryland 20706
www.rowman.com

Unit A, Whitacre Mews, 26-34 Stannary Street, London SE11 4AB

British Library Cataloguing in Publication Information Available

Library of Congress Cataloging-in-Publication Data Available

ISBN 978-1-4985-3569-4 (cloth : alk. paper)
ISBN 978-1-4985-3570-0 (electronic)

∞™ The paper used in this publication meets the minimum requirements of American National Standard for Information Sciences—Permanence of Paper for Printed Library Materials, ANSI/NISO Z39.48-1992.

Printed in the United States of America

Contents

Acknowledgements vii

Introduction: Ecological Crises, Nonhumans and the Age of Man ix
Sy Taffel and Nicholas Holm

PART I: NONHUMAN AGENCY **1**

1 Carbon Bonds: Coal Economics and Aesthetics 3
Sean Cubitt

2 Consider the Lawnmower: Aesthetics, Politics and Entanglements of Suburban Nature 17
Nicholas Holm

3 Learning with the River: On Intercultural Gifts from the Whanganui 35
Charles Dawson

4 From Wai 262 to Water: Towards Postcolonial Property Rights in Aotearoa New Zealand 55
Jacob Otter

PART II: CULTIVATION AND CULTURE **71**

5 The Plough as Settler Colonial Cultural Icon: Voices from the *Other* Side of the Blade 73
Victoria Grieves

6 Conserving Land through Kindly Use and Reciprocity: Using the Land and Being Used by the Land 91
Anne O'Brien

7 "One Loaf of Bread at a Time": System Change through Community Food Initiatives 107
Sharon Stevens

8 In Different Voices: Engaging with Human–Nonhuman Entanglements 125
Sita Venkateswar

PART III: EPISTEMOLOGY, AESTHETICS AND MEDIATION **141**

9 Photographic Reflections on Landscape Change in Regional Australia 143
Christopher Orchard and James Holcombe

10 Nature as Creative Catalyst: Building Poetic Environmental Narratives through the Artists in Antarctica Programme 159
Octavia Cade

11 Mediating the "Deep": A Partial Genealogy of Media Work with Oceans and Seas 177
Gareth Stanton

12 Exiled in the Bush: A History of Landscape Transformation in Post-European Settlement New South Wales 197
David Orchard and Peter Orchard

13 Mapping the Anthropocene 219
Sy Taffel

Index 241

About the Contributors 247

Acknowledgements

This anthology began its life as a conference, "Working with Nature: Ecological Entanglements in the Anthropocene," which was held in April 2015 at the Manawatu campus of Massey University in Aotearoa New Zealand. Initially envisioned as a small-scale symposium event, we were taken aback by the high levels of interest among those eager to discuss and debate the implications of the Anthropocenic moment, and our small event gradually shifted to a larger two-day format that provided a forum for a range of voices and perspectives addressing the different ways in what we once thought of as natural and human worlds might relate to one another. This book would never have existed without both the support of various institutions and the enthusiasm of the local academic community which made that event possible.

We would therefore like to thank Massey University's School of English and Media Studies, College of Humanities and Social Sciences and, especially, the W.H. Oliver Humanities Research Academy (and its director, Kerry Taylor) for their generous support of the initial event. After the fact, the Massey University Political Ecology Research Centre also provided support that helped make this book possible.

We would also like to thank all those who presented at the conference – Ozan Alakavuklar, Cameron Boyle, Sean Cubitt, Charles Dawson, Caroline Depatie, Ali East, Victoria Grieves, Dinah Hawken, Susie Lachal, Jenny Mauger, Sasha Matthewman, Anne O'Brien, Jacob Otter, Christopher Orchard, David Orchard, Peter Orchard, Kathryn Ovenden, Tamati Patuwai, Jane Richardson, Isa Ritchie, Janet Sayers, Heike Schiele, Gareth Stanton, Polly Stupples, Sita Venkateswar, Chris Teo-Sherrell and Helen Lehndorf – who all contributed to the discussions and debates that shape the arguments and perspectives presented in this collection. In addition, the assistance of

Carolyn Morris, Ingrid Horrocks and Tim Jurgens in the organization of the conference helped make the event possible.

Finally, we would also like to thank those who helped make this collection possible: Thom Conroy stepped in to help us locate the right person to fill a gap that opened up at the last second; Lexington Books' Ecocritical Theory and Practice series editor, Douglas Vakoch, who first approached us with the idea that the conference might serve as a good basis for an anthology; and our editor Lindsey Porambo, who provided invaluable advice and support as we prepared this collection.

Introduction

Ecological Crises, Nonhumans and the Age of Man

Sy Taffel and Nicholas Holm

In the twenty-first century, it is becoming increasingly apparent that whether we consciously address it or not, human cultures and societies are entangled with what was once called nature. Nature is one of those ever-present yet somewhat uncomfortable words that structure our everyday lives. Even as we acknowledge that the concept carries troublesome philosophical baggage, nature persists as an almost inescapable term of reference in discussions of environmental and ecological concerns. Entanglement, as outlined by the philosopher Karen Barad, is a useful concept in addressing these matters, insofar as it posits that phenomena cannot be considered in isolation from one another.[1] Entanglement thus departs from terms such as interconnection and networking, which suggest that existence is composed of entities that are connected, but remain functionally separable from one another. Rather than a world populated by already-formed individual agents which interact with one another, entanglement instead posits existence as an unfolding process in which flows of energy and matter cannot be neatly divided into segregated agents who occasionally come together. As opposed to a world where categorically distinct entities – nature and cultures, subjects and objects, or even humans and nonhumans – enter into conflict or concord with one another, entanglement instead suggests that systems must always be conceptualized within collective terms which denotes that existence as a process of acting together. This acting together is inherently a process of producing differences within the world, what Barad describes as the performative generation of diffraction patterns throughout entangled systems.[2] While the technical discussion of entanglement in Barad's work stems from an in-depth engagement with quantum physics, and in particular the work of Niels Bohr, the manner in which the term is employed by the diverse interdisciplinary essays within this book does not strictly follow the onto-epistemology laid out within

Barad's work, but are broadly influenced by the conception of entanglement as a mode of collectively becoming-together which goes beyond discourses of connectivity and interaction.

ENTANGLEMENT AND THE ANTHROPOCENE

Entanglement therefore parallels certain aspects of ecology, the study of flows of energy and matter through multiscalar systems encompassing individuals, populations (groups of the same species), communities (agglomerations of multiple species) and ecosystems (which explores non-living nonhumans in addition to communities of life).[3] Whereas the term "environment" derives from the Old French term "to surround," suggesting something which is fundamentally external to the subject, ecology instead presents an immanent field in which multiple forms of living and non-living beings converge to form dynamic and complex systems situated far from a point of entropic equilibrium, but capable of allowing life to continue reproducing itself. From the perspectives offered by both ecology and entanglement, "nature" has never been something outside of humans and culture, an exterior environment which provides the raw materials for cultural innovation. Ecological ontologies thus resonate with the monistic world views advanced by in recent years by posthumanists,[4] Actor-Network Theorists,[5] new materialists,[6] and object-orientated ontologists.[7] Rather than the Cartesian demarcation between nature and culture, these otherwise disparate perspectives share the rejection of a dualistic ontology which separates humans from nature and brackets the latter term as an environment for social action.

Such a manner of conceptualizing the relation between those categories formerly known as nature and culture becomes particularly pertinent in the current moment of the Anthropocene. A literal translation of the Anthropocene combines the suffix -cene, which denotes an epoch situated within the Cenozoic era, alongside the prefix Anthropos which derives from the Greek word "ἄνθρωπος," which has been traditionally translated by biblical scholars as "Son of Man."[8] While certain contemporary applications of this prefix refer to study of humans in a non-gender-specific manner, such as within the field of anthropology, we should be wary of erasing the gendered origins of this usage, which initially referenced the "the science of Man."[9] Definitions of the term "Anthropocene" consequently vary from Paul Crutzen's initial discussion of a "Geology of Mankind," through Elizabeth Kolbert's "Age of Man," to broader variants which suggest an age of humans or humankind.[10] In this introduction we have chosen to adopt Kolbert's formulation, in part due to its productive resonance with Michel Foucault's oft-cited statement drawn from the conclusion of *The Order of Things*, that "Man is an invention of recent date. And one perhaps nearing its end."[11] As we shall see, while humanity often refers to a somewhat nebulous and highly problematic universalization which homogenizes a range of local, embedded and specific human actions, the ecological and geological contexts gestured towards by the Anthropocene are closely tied to the universalist ideology of the enlightenment subject whose historical invention Foucault identifies, and the economic model of industrial capitalism which accompanied it.

In recent years, the discourse of the Anthropocene has been increasingly employed in scientific, critical and popular contexts to mark a new unit of geological time which denotes that human interventions and alterations of the nonhuman world have created global conditions which markedly and measurably depart from those of the Holocene: the previous epoch within the geological time scale. The Holocene began 11,700 years ago at the onset of the current interglacial period, where thawing permafrost resulted in a sea level rise of over a hundred metres. While human beings had already migrated to every continent on Earth spare Antarctica before the onset of the Holocene, and have undertaken notable processes of ecosystem engineering with widespread ecological impacts for thousands of years,[12] the Anthropocene is said to signal change at different spatio-temporal scales to these historical events.

One of the key markers of this new era is the unassailable evidence for human-driven climate change,[13] which is frequently presented as a paradigmatic example of such alterations. Atmospheric concentrations of carbon dioxide have risen from approximately 280 ppm before the industrial revolution in 1880, to over 400 ppm today,[14] concentrations not seen in the 800,000 years of ice core data from which humans can reliably read levels of greenhouse gases. Rising carbon dioxide levels are a global phenomenon, whose various and variable effects will persist for time scales which dwarf those of human civilizations, let alone the lives of individual humans, and the impacts of a warmer world, including sea level rise which is predicted to redraw coastal regions, flooding deltas and low-lying cities, will be visible in the archives of the Earth itself such as polar ice cores for longer still.

Climate change is not the only geological evidence gestured towards to support the thesis of the Anthropocene; the concept additionally describes wider transformations regarding the entanglement of humans and nonhumans across numerous ecological contexts. The last century has seen the production of vast quantities of non-biodegradable materials which will be present in some form within the geological strata of the planet for millions of years. Approximately 98 percent of the purified, elemental aluminum in existence, over 500 million tons in mass has been produced since 1950.[15] Similarly, over 300 million tons of synthetic, petrochemical-derived plastics are produced annually, with no synthetic polymers having been produced before 1900.[16] These technofossils will provide a relatively sudden change within stratigraphic records, as while humans have long produced forms of biologically persistent technological materials such as glass, the scale of globalized industrial production associated with the "great acceleration" post-Second World War produced the level of geological evidence necessary for denoting epochal change. Further examples of the unprecedented intrusion of the human into the realm previously known as nature include the increased rate of species extinction, with recent data suggesting that contemporary extinction rates for vertebrates are up to one hundred times higher than background extinction rates, and have risen over the course of the past century[17] the anthropogenic global movements of species associated with agricultural practices; the widespread transformation of ecosystems to support intensive agricultural activity and industrial fishing practices[18]; and alterations to planetary ecological processes such as the nitrogen cycle, where anthropogenic nitrogen fixing is argued to have produced the "the largest impact on the nitrogen cycle since the major pathways of the modern cycle originated some 2.5 billion years ago."[19]

The most abrupt, globally synchronous evidence for the Anthropocene is the presence of radionuclides which result from the anthropogenic usage of nuclear weapons. These devices leave enduring traces of their destructive impacts in the form of isotopes which are rarely found otherwise, such as plutonium-239, whose half-life is over 24,000 years, entailing that this precise dating of the Anthropocene will be detectible in sediments and ice cores for in excess of 100,000 years.[20] Symbolically, we can perhaps note the manner that the Anthropocene's rhetorics of ecocide and the inadvertent anthropogenic destruction of Holocenic planetary systems dovetail here with a stratigraphic marker demarcating the pinnacle of human-led technological progress in the deliberate slaughter of life via weapons of mass destruction.

FROM PLANETARY TO LOCAL ANTHROPOCENE

The Anthropocene does not, however, merely suggest that humans are altering Earth's ecosystems. If this were the case, then the Anthropocene would be a period coterminous with the existence of our species, but such a periodization would represent a hubristic anthropocentricism which fails to acknowledge the ways that a multitude of other species also affect ecosystems across the globe. To take one example, beavers are important ecosystem engineers who modify the flow of rivers through the constructions of dams, thereby altering the ability of numerous other species within freshwater ecosystems to swim, spawn, hunt, hatch, hide and otherwise dwell within the locale populated by the beaver, while many of these organisms, such as predators, prey, symbionts, parasites and commensuals will feedback into the continuing existence of the beavers. This foregrounds the manner by which ecological entanglement highlights the complexity and dynamism, of mutable ecosystems whereby there is no exterior being affected by active agents; the entangled entities of ecosystems form iterative patterns of cause and effect. By virtue of being part of an ecosystem an entity has innumerable impacts upon the range of entangled entities which collaboratively constitute that ecology.

Proponents of the Anthropocene consequently argue that the epochal boundary is formed not by visible changes to a locale, but by the sum impact of humanity creating changes which act upon the Earth System at a global scale in a manner which will have significant consequences for durations measured in hundreds of thousands, or millions of years, and that will be visible as alterations to the formation of geological strata.[21] The conception of the Earth System being advanced here is a relatively new interdisciplinary approach which emerges from the Earth sciences, within which "the *Earth System* has come to mean the suite of interacting physical, chemical, and biological global-scale cycles (often called biogeochemical cycles) and energy fluxes which provide the conditions necessary for life on the planet,"[22] From this perspective, it is argued that human action is itself a component of the Earth System, and that:

> Human activities have become so pervasive and profound that they rival the great forces of Nature and are pushing the Earth into planetary *terra incognita*. The Earth is rapidly moving into a less biologically diverse, less forested, much warmer, and probably wetter and stormier state.[23]

Although forecasting complex and dynamical systems such as the global climate predicts a broad range of possible futures depending upon both future anthropogenic activities and non-anthropogenic forgings and feedbacks, the simulations produced by global climate models indicate that as we progress through the twenty-first century, life on Earth will not only become significantly harder for many humans due to the changing climate, but that we will very likely see the sixth planetary mass extinction event write itself into the geological archives of the planet.[24] The most recent great dying of life on the planet, the Cretaceous-Paleogene extinction event, which saw the loss of approximately 75 percent of all plant and animal species – including, most infamously, the dinosaurs – was brought about by an asteroid impacting the planet. The forecast for the Anthropocene suggests that today humanity has become a similar force of death; our apparently endless thirst for consumption has seen us become the biological equivalent of an asteroid, whose collision with Earth's ecosystems will leave both the asteroid and the planet itself radically altered.

While we may query the degree of separation posited by such accounts of humanity acting upon a rather passive and fragile world, and the accompanying claims that we must act to "save the planet," we should acknowledge that on a global scale, we are witnessing an increasing concern with the different ways in which human behaviour works to shape nature. From climate change, and global reductions in biodiversity to drives towards sustainable communities and ongoing concerns with waste and pollution, entanglements between humans and nonhumans looks set to be a central concern of the twenty-first century. However, even as recognition of anthropogenic climate change and the specter of a (hu)man(kind)-made mass extinction event become commonplace, and concepts such as sustainability and resilience enter into the conversations of state and corporate actors, it remains unclear how these ideas might speak to our everyday practices and behaviour, or how global models of the Anthropocene can impact upon on national, regional and local communities and social structures.

Even as the notion of the Anthropocene gains increasing currency in scientific, journalistic and popular discourse, the revelation of humanity's inextricable implication in the nonhuman world remains overwhelmingly orientated towards a grand scale. Climate change is a prime example of this tendency whereby the Anthropocene provides the intellectual context by which to imagine collective human agency on a planetary level. Rather than considering how the nonhuman interpenetrates and influences multiple aspects of everyday life, or explicating the heterogeneous impacts of different groups of humans, heralds of the Anthropocene frequently seem more concerned with how humanity's agency expands out to match the largest manifestations of the nonhuman world, such as Steffen et al.'s invocation of the "Great Forces of Nature."[25] Operating on this scale, the stakes of human action appear terrifyingly high. This is not surprising: after all, given the damning litany of degradations that humanity has visited upon the nonhuman world there are good reasons to fear the consequences of human activity writ large. However, in this concern with the big picture, Anthropocenic thinking threatens to perpetuate the assumption that it was meant to finally overturn: the belief that the natural or nonhuman world is something *out there*. The only difference is that in the Anthropocene, we as a species are now *out there* as well. What is often overlooked in these grand imaginings are the more every day and intimate entanglements of the human and nonhuman that are just as much a part of Anthropocenic existence. At the intimate level of the constitution of the corporeal bodies we identify as "humans," we are of course deeply enmeshed with forms of life that would ordinarily be classified as nonhuman. Not only would our bodies be unable to extract sufficient energy from food to us to remain alive without the microbiota which dwell within our intestines, but the ecology of life within the boundaries of our skin has recently been demonstrated to contain ten times as many bacterial cells as "human" cells.[26] Taking such evidence into account requires rethinking the notions of being human and human beings, departing from the Cartesian model of the unified and singular subject, instead positing large organisms as dynamic mutualistic multispecies assemblages. On a related note, the entanglements between humans, food crops, soils, pests, pollinators, transportation and harvesting technologies and socio-economic structures among other entities are, as Sharon Stevens' chapter in this volume foregrounds, evolutionary systems which act as collective becomings-together. What we eat, and how food is produced has a series of cascading consequences for these systems which demonstrates that while the Anthropocene often projects humans *out there* seeping into what used to be called nature, nonhumans must equally be understood as always *in here*, constituting the unfolding processes that we designate as being human.

There are, however, good reasons as to why the Anthropocene is usually discussed on a planetary, rather than local, scale. In its strictest sense, the concept operates on the levels of geological time – a chronological scale measured in hundreds of millions of years that renders the two hundred years since the industrial revolution to be less than a blink of the eye – and globalized space as apertures for exploring entanglements within the Earth System. At such vast scales the nation, let alone a single plot of private land, matters little. However, despite the Anthropocene always gesturing towards global space and geological time, we should note that the entangled intra-actions which compose this novel geological epoch are constituted by a myriad of heterogeneous actions which take place in the differentiated spaces of globalized capitalism and which occur at speeds measured in days, weeks or years rather than millennia.

If we are focused only on our immediate contexts, then for most of us, there is little possibility of debate that our environments are almost entirely the product of directed and long-standing human action and have been this way for centuries if not millennia. This applies not only to the obvious examples such as cities and farms, but to the immense long-standing projects of landscape transformation that have been carried out around the world for centuries, from river diversion to hillside terracing. The observation that the nonhuman is inextricably shaped by the human is therefore not a revelation at these smaller scales: it is simply a fact of life. Nonetheless, even though the entanglement of human and nonhuman worlds is less surprising on smaller scales, this does not mean that processes operating on that level are entirely divorced from the context of the planetary Anthropocene. As individuals, or even communities, we do not live on a planetary scale: we live in a local environment. However, it is the sum total of those heterogeneous actions taken at local levels that have given rise to the Anthropocene. The larger entanglements that define this new epoch can thus be thought as the result of the accumulation of many smaller acts of extraction, combustion, predation, production and consumption. We must, however, emphasize that whereas the Anthropocene's headline of the Age of Man subsumes and homogenizes a diverse array of actions, when we seek to mobilize, alter, affect, impact or otherwise engage with the consequences of contemporary ecological crises, we can only do so as embodied actants situated at particular points within space and time. If conceptualizing the sum of human changes to global systems asks us to go from the local to the planetary Anthropocene, then taking action requires us to return to the here and now.

WORKING WITH NATURE

The Anthropocenic moment is consequently more than simply a matter of scientific evidence and definition, but instead represents a profound transformation in some of the most basic conceptions of human capability and responsibility. In order to meet the challenges and possibilities of this new moment, it is therefore essential that the questions posed by the Anthropocene are not conceived solely in scientific terms, but also account for the key ethical, aesthetic, political and social issues surrounding contemporary entanglements between human and nonhuman systems. Acknowledging the important and rapidly developing conversations already taking place with regard to these issues, *Ecological Entanglements in the Anthropocene* seeks to contribute to this ongoing discussion through a range of transdisciplinary interventions exploring what it might mean to conceive of these ecological entanglements and political issues in terms of "working with nature." The essays collected in this volume are selected from material originally presented at the "Working with Nature" conference hosted at Massey University, Aotearoa New Zealand in March 2015. The conference brought together activists, artists, and a range of academic contributors from across multiple disciplines to examine how the idea of "working with nature" might help us understand and assess the different ways in which humans transform, engage and interact with the nonhuman world. Over two days, delegates engaged with a series of interrelated questions in response to the conference's central provocation that asked what would it mean to regard "nature" as something that can be worked with: Are there more or less preferable ways of working with nature? What forms and practices might this work take and how do we distinguish between them? Is the idea of "nature" even sufficient to approach such questions, or do we need to reconsider using the term nature in favour of terms such as environments, ecologies or the broad notion of the nonhuman world?

At the heart of these discussions, and the essays collected in this anthology, is the idea that working *with* nature provides the possibility for new conversations about ecological entanglements. In particular, working *with* nature needs to be distinguished from working *for* or *against* nature: different orientations that have tended to dominate ecological thinking in the twentieth and early twenty-first centuries. On the one hand, the concept of working *for* nature has tended to be the driving force of environmental, and specifically conservationist and deep ecological, orientations towards the nonhuman world. To work for nature means to put oneself in service to it: to orientate oneself towards life as an entity that is schizophrenically and reductively conceptualized as simultaneously being our master and a fragile charity case. When we work for nature, we tend to presume that not only do we know what is best for it, but that even more fundamentally we know what nature is. Such a position is thus simultaneously potentially generous in its willingness to put nature's interests before our own, but also highly presumptive in its assumption that we can determine what is best for nature and act in its interests, indeed even the notion that there can be a singular path which benefits nature in its entirety is one which cannot bear much scrutiny. On the other hand, while few would claim to be working *against* nature, the narrative implied in such a formulation is one of the consistent recurring themes of environmental discourse. Polluters, extractors, emitters, habitat destroyers: these subjects are seen to be in the wrong because they are performing actions that are not in the interest of the nonhuman world. To work against nature is more than simply to be in conflict with a particular set of forces, processes and actors; it is to act in an unnatural way. Here the meaning of the unnatural slides into ethical and aesthetic registers as it conjures up ways of being and becoming that are not just deleterious for particular living entities, but – courtesy of the capricious meanings of "nature" – are also somehow opposed to the fundamental way that the world should be arranged. Thus if to work for nature is to be arrogantly unreflexive; to work against nature is to be profoundly and completely incorrect. Neither option would seem particularly desirable.

The idea of working with nature is thus conceived to rescue us from this impasse, by engaging with nature as something that need not be known in any certain sense. To work either for or against nature, one needs to have some sense of nature's ends: both "for" and "against" imply a knowledge of nature's intention. There is no such assumption when we work *with* nature: only an awareness that the human is not the only site of agency, action and influence in this encounter. Consequently, when we conceive of human-nonhuman entanglements in terms of working with nature, there is no longer a need to presume what exactly nature is, or what precise forms it does or should take. Instead, to work with nature means to acknowledge both that nonhumans have agencies – that they exert influences, pressures and set limitations upon human activities – and that it is possible for entangled entities to act upon one another without damage necessarily being done to the participants. To work with nature is thus also to begin to venture beyond the concept of nature and all the provocation and problems that it carries within itself and to thereby enter into that philosophical and theoretical space that exists beyond or after nature.[27] The idea of "working with nature" is thus also particularly well-suited to our current moment of the Anthropocene, when that always blurry distinction between nature and non-nature begins to collapse altogether. Under such conditions, when it is no longer possible to gesture towards a terrestrial or oceanic wilderness which exists beyond the realm of human ecosystem engineering, and where we are confronted with the knowledge that microbes living within the confines of our skins outnumber our somatic cells tenfold,[28] it becomes inevitable that whether we seek to or not, whether we desire to or not, we will always be working with nature.

SECTIONS AND CHAPTERS

In ordering diverse, interdisciplinary interventions, this anthology is focused around three key themes that help emphasize the novel and necessary contributions of the individual essays: nonhuman agencies, cultivation and culture, and epistemologies, aesthetics and mediation. The first section, nonhuman agencies, comprises essays which address different ways that various agencies associated with heterogeneous forms of nonhumans come to form pivotal parts of contemporary social, cultural and technological forms. In "Carbon Bonds: Coal Economics and Aesthetics," Sean Cubitt explores the way that the past, present and future of audiovisual media are entangled with coal. From the first synthetic inks used to print books and newspapers through to contemporary carbon futures markets which are a vital component of globalized financial capitalism, Cubitt maps the way that humans have transformed the fossilized remains of long-dead plants into a multitude of forms that have forged the conjuncture currently described as the Anthropocene. From a markedly different perspective, Nicholas Holm's chapter, "Consider the Lawnmower: Aesthetics, Politics and Entanglements of Suburban Nature," then reconsiders suburban lawn mowing practices as political and aesthetic acts which gestures towards ways that human engagements with living nonhumans can move beyond an oppositional rhetoric of mastery and domination. Holm challenges claims that lawn mowing is a homogeneous act aimed at controlling nature by exploring the local and variable contexts, practices and nonhumans associated with lawn mowing, providing a model for how we may productively engage with the global discourse of the Anthropocene at a hyperlocal scale while retaining a productive sense of how human agency can shape nonhumans.

This first section also features two chapters with a particular focus upon the value systems and property rights of Māori, the indigenous people of Aotearoa New Zealand, particularly with respect to rivers and other forms of water, thus providing an important alternative to Western models of ownership based upon privatization and liberal rights of the individual. Traditional Māori knowledge conceives of the nonhuman world as a site of centrally important agency and influence, and both chapters draw on that knowledge as a means to articulate an alternative ontology. Charles Dawson's chapter, "Learning with the River: On Intercultural Gifts from the Whanganui" presents a personal history narrative surrounding the role of his father in New Zealand's Treaty of Waitangi settlement claims process as a means of outlining Māori ontological claims regarding the agencies of water and other entities which have been traditionally relegated to the roles of passive objects by Western legal frameworks. Bringing together personal reminiscence, postcolonial theory and ongoing legal conflicts in Aotearoa New Zealand, Dawson articulates a politically and ethically engaged vision of embodied entanglement. In a related argument, Jacob Otter examines the Treaty of Waitangi Wai 262 claim surrounding state-led claims of ownership of the Whanganui River and attempts to sell this resource to a private company for hydroelectric power generation. In his chapter, "From Wai 262 to Water: Towards Postcolonial Property Rights in Aotearoa New Zealand," Otter takes up this battle of property rights as a way to consider the ramifications of Māori ontology for dominant colonial conceptions of the nature-culture distinction and the pragmatic legal ramifications that follow from that.

The essays in the second section, Cultivation and Culture, address the forms of human–nonhuman entanglement found within agricultural practices and food production, a vital element within our relationships to the nonhuman world, but one whose material practices and cultural consequences are relatively underexplored within existing ecocritical debates. Consequently, the chapters within this section examine a range of agricultural practices, taking differentiated approaches to considering how ethics of care are contained within practices of cultivation, land transformation and crop growth, forms of human–nonhuman relationship which are often eschewed by a focus upon charismatic megafauna and notions of nature as wilderness, a romantic other which sits outside human practices of culture and cultivation. Victoria Grieves' chapter "The Plough as Settler Colonial Cultural Icon: Voices from the *Other* Side of the Blade," examines technological developments and social practices surrounding the mechanization of the plough and its deployment in the colonial contexts of Australia, Aotearoa New Zealand and North America. Grieves argues that the plough exemplifies a logic based upon geographical and cultural homogenization allied with a misguided belief in human mastery over nature, which she identifies as being central to the onset of contemporary ecological crises. In "Conserving Land through Kindly Use and Reciprocity: Using the Land and Being Used by the Land," Anne O'Brien cuts to the core of what it might mean to engage with the nonhuman world in a mutually productive way. By taking up the concept of kindly use, O'Brien advocates for a form of ecological stewardship and care that draws upon both indigenous knowledge and European phenomenology. O'Brien illustrates this idea of kindly use through case studies of soil management in contemporary Australia. Similarly, Sharon Stevens is concerned with the wider lessons that can be learnt from local landcare practices. Drawing upon her own work as a community activist and organizer, alongside interviews with numerous grass-roots community groups in the Manawatu region of Aotearoa New Zealand, Stevens presents perspectives on how and why those involved in community food initiatives are sufficiently moved to invest time and energy into such efforts. This chapter, "'One Loaf of Bread at a Time': System Change through Community Food Initiatives," explores how place-based community food activism transforms ecological relationships within specific places, and how this focus on the local contrasts with the homogenized logic of monocultural industrialized food production. Finally, Sita Venkateswar's chapter, "In Different Voices: Engaging with Human–Nonhuman Entanglements," combines two strands of related enquiry into assemblages of life which always exceed the domain of the human. Venkateswar recounts narratives from her family history, recalling ways of living which emphasize cohabitation rather than a misguided sense of human mastery over nature, before exploring how the production of millets within contemporary India gesture towards a mode of agricultural production which escapes the monocultural methods of intensive approaches, instead forming resilient multispecies meshworks which are mutually beneficial to human and nonhuman actors.

The final section of the book, entitled Epistemologies, Aesthetics and Mediation, features a range of positions which explore how humans represent the nonhuman world in order to mobilize particular comprehensions of ecological entanglement which are in turn designed to stimulate collective activities that intervene in and engage with ecological crises. While traditional notions of conservation suggest the preservation of an idealized pre-human state of nature, this section instead seeks to address how representing the nonhuman requires an engagement with the turbulence and dynamism of nature. The authors in this section additionally ask us to consider the manner through which many of our collective understandings of nature, whether as wilderness, global ecological crises, or historical transformation, primarily involve engaging with mediated representations rather than directly experiencing phenomena and consequently ask what kinds of aesthetic strategies are likely to be successful here in terms of affecting change and generating a collective sense of care for future generations of humans and nonhumans. In "Photographic Reflections on Landscape Change in Regional Australia" Christopher Orchard and James Holcombe critically engage with the traditions of landscape photography and its ties to national identity. Utilizing concepts drawn from ecopsychology to explore how photographic practice can encourage the embrace of an environmental mutualism, Orchard and Holcombe consider how our encounters with photographic images can lead us to reimagine landscape and our relations to the nonhuman world. Operating from a different perspective, Octavia Cade then asks an associated question: What kind of knowledge can we gain from aesthetic encounters with places and landscapes that we are never likely to actually experience? Through an investigation of poetry addressing Antarctica, Cade's chapter, "Nature as Creative Catalyst: Building Poetic Environmental Narratives through the Artists in Antarctica Programme," argues that when environments are unable to be experienced directly, aesthetic forms, such as poetry, become an important counterpoint to dominant scientific narratives.

While the vast majority of ecocritical work focuses upon the terrestrial fraction of our blue planet, Gareth Stanton's chapter, "Mediating the 'Deep': A Partial Genealogy of Media Work with Oceans and Seas," considers the embodied and imagined relationships between humans and Earth's oceans. Stanton explores how human comprehensions of the oceanic depths and the creatures who dwell there are primarily shaped by various representational media, ranging from the novels of the nineteenth century through to contemporary documentary films which detail the immense slaughter of life in the oceans, but which have had seemingly little effect on humanity's insatiable appetite for eating fish. David and Peter Orchard's chapter, "Exiled in the Bush: A History of Landscape Transformation in Post-European Settlement New South Wales," then explores multiple ways that changing attitudes towards Australian environmental history have been driven by economic, social and cultural factors. The Orchards argue that uncovering how material practices are shaped by epistemological constructions forms a crucial precondition to taking effective action to address contemporary ecological crises such as climate change and declining biodiversity. Finally, in "Mapping the Anthropocene," Sy Taffel examines the intersections between the geological debates surrounding how to best quantify and classify the Anthropocene, the computational and simulational technologies which are pivotal to the production of these scientific discourses, and cultural critiques of the Anthropocene which highlight the problematic reduction of a diverse assemblage of differential human and nonhuman actors under the banner of the Age of Man. Taken together the different sections and chapters of this anthology thus represent a range of possible answers, arguments and avenues for exploration regarding the inescapable ecological entanglements that will come to increasingly define what it means to be human in the Anthropocene.

NOTES

1. Karen Barad, *Meeting the Universe Halfway: Quantum Physics and the Entanglement of Matter and Meaning* (Durham and London: Duke University Press, 2007).

2. Barad, *Meeting the Universe Halfway*, 89.

3. Michael Begon, Colin R. Townsend, and John L. Harper, *Ecology: From Individuals to Ecosystems* (Malden, MA: Blackwell, 2006).

4. Rosi Braidotti, *The Posthuman* (Cambridge: Polity, 2013); Donna Haraway, *The Companion Species Manifesto: Dogs, People, and Significant Otherness* (Chicago: Prickly Paradigm Press, 2003); Cary Wolfe, *What Is Posthumanism?* (Minneapolis: University of Minnesota Press, 2010).

5. Bruno Latour, *Politics of Nature: How to Bring the Sciences into Democracy* (Cambridge, MA: Harvard University Press, 2004); John Law, "After Ant: Complexity, Naming and Topology," *The Sociological Review* 46, no. S (1998): 1–14.

6. Jane Bennett, *Vibrant Matter: A Political Ecology of Things* (London: Duke University Press, 2009); Diana Coole and Samantha Frost, *New Materialisms: Ontology, Agency, and Politics* (Durham: Duke University Press, 2010); Gilles Deleuze and Félix Guattari, *A Thousand Plateaus: Capitalism and Schizophrenia* (Minneapolis: University of Minnesota Press, 1987); Gilles Deleuze and Félix Guattari, *Anti-Oedipus: Capitalism and Schizophrenia* (New York: Viking Press, 1977).

7. Levi Bryant, *Onto-Cartography: An Ontology of Machines and Media* (Edinburgh: Edinburgh University Press, 2014); Timothy Morton, *Ecology without Nature: Rethinking Environmental Aesthetics* (Cambridge, MA: Harvard University Press, 2007).

8. Delbert Burkett, *The Son of Man Debate: A History and Evaluation* (Cambridge: Cambridge University Press, 2000); Carl Hermann Kraeling, *Anthropos and Son of Man* (New York: Columbia University Press, 1927).

9. James Hunt, "Introductory Address on the Study of Anthropology," *The Anthropological Review* 1 (1863): 1–3.

10. Paul J. Crutzen, "Geology of Mankind," *Nature* 415, no. 6867 (2002): 23; Kolbert, Elizabeth, "Enter the Anthropocene – Age of Man," *National Geographic* 219, no. 3 (2011): 60; Jan Zalasiewicz, Mark Williams, Will Steffen, and Paul Crutzen, "The New World of the Anthropocene," *Environmental Science & Technology* 44, no. 7 (04/01/2010): 2228–31.

11. Michel Foucault, *The Order of Things: An Archaeology of the Human Sciences* (London and New York: Routledge, 2002).

12. William F. Ruddiman, "The Early Anthropogenic Hypothesis: Challenges and Responses," *Reviews of Geophysics* 45, no. 4 (2007): 1–37; William F. Ruddiman, "The Anthropocene," *Annual Review of Earth and Planetary Sciences* 41, no. 1 (2013): 45–68.

13. While the language of climate change has a political connotation in the United States which relates to the George W. Bush administration's adoption of the term in favour of global warming in an attempt to systematically downplay the seriousness of the issue, the formation of the Intergovernmental Panel on Climate Change in 1988

demonstrates that the term "climate change" significantly predates the Bush administration, and has long been the preferred terminology within scientific discourse.

14. Damian Carrington, "Global Carbon Dioxide in Atmosphere Passes Milestone Level," *The Guardian*, 10th May 2013, https://www.theguardian.com/environment/2013/may/10/carbon-dioxide-highest-level-greenhouse-gas.

15. Jan Zalasiewicz et al., "The Technofossil Record of Humans," *The Anthropocene Review* 1, no. 1 (2014): 34–43.

16. Gaelle Gourmelon, *Book Global Plastic Production Rises, Recycling Lags* (Washington, DC: Worldwatch Institute, 2015).

17. Gerardo Ceballos et al., "Accelerated Modern Human – Induced Species Losses: Entering the Sixth Mass Extinction," *Science Advances* 1, no. 5 (2015): 1–5.

18. Erle C. Ellis et al., "Anthropogenic Transformation of the Biomes, 1700 to 2000," *Global Ecology and Biogeography* 19, no. 5 (2010): 589–606.

19. Donald E. Canfield, Alexander N. Glazer, and Paul G. Falkowski, "The Evolution and Future of Earth's Nitrogen Cycle," *Science* 330, no. 6001 (2010): 192–6.

20. Colin N. Waters et al., "Can Nuclear Weapons Fallout Mark the Beginning of the Anthropocene Epoch?" *Bulletin of the Atomic Scientists* 71, no. 3 (2015): 46–57.

21. Clive Hamilton, "Getting the Anthropocene So Wrong," *The Anthropocene Review* 2, no. 2 (2015): 102–7; Colin N. Waters et al., "A Stratigraphical Basis for the Anthropocene?" *Geological Society, London, Special Publications* 395, no. 1 (2014): 1–21; Jan Zalasiewicz et al., "When Did the Anthropocene Begin? A Mid-Twentieth Century Boundary Level Is Stratigraphically Optimal," *Quaternary International* 383 (2015): 196–203.

22. Will Steffen, Regina Angelina Sanderson, Peter D. Tyson, Jill Jager, Pamela Matson, Berrien Moore III, Frank Oldfield, Katherine Richardson, Hans Joachim Schellnhuber, and Billie L Turner, *Global Change and the Earth System: A Planet under Pressure* (Berlin: Springer, 2004).

23. Will Steffen, Paul J. Crutzen, and John R. McNeill, "The Anthropocene: Are Humans Now Overwhelming the Great Forces of Nature," *AMBIO: A Journal of the Human Environment* 36, no. 8 (2007): 614.

24. Elizabeth Kolbert, *The Sixth Extinction: An Unnatural History* (London: A&C Black, 2014).

25. Steffen et al., *AMBIO*, 2007.

26. Peter J. Turnbaugh, Ruth E. Ley, Micah Hamady, Claire M. Fraser-Liggett, Rob Knight, and Jeffrey I. Gordon, "The Human Microbiome Project," *Nature* 449, no. 7164 (2007): 804–10.

27. Morton, *Ecology without Nature*; Jedidiah Purdy, *After Nature* (Cambridge: Harvard University Press, 2015).

28. Turnbaugh et al., *Nature*, 2007.

BIBLIOGRAPHY

Barad, Karen. *Meeting the Universe Halfway: Quantum Physics and the Entanglement of Matter and Meaning*. Durham and London: Duke University Press, 2007.

Begon, Michael, Colin R. Townsend, and John L. Harper. *Ecology: From Individuals to Ecosystems*. Malden, MA: Blackwell, 2006.

Bennett, Jane. *Vibrant Matter: A Political Ecology of Things*. London: Duke University Press, 2009.

Braidotti, Rosi. *The Posthuman*. Cambridge: Polity, 2013.

Bryant, Levi. *Onto-Cartography: An Ontology of Machines and Media*. Edinburgh: Edinburgh University Press, 2014.

Burkett, Delbert. *The Son of Man Debate: A History and Evaluation*. Cambridge: Cambridge University Press, 2000.

Canfield, Donald E., Alexander N. Glazer, and Paul G. Falkowski. "The Evolution and Future of Earth's Nitrogen Cycle." *Science* 330:6001 (2010): 192–6.

Carrington, Damian. "Global Carbon Dioxide in Atmosphere Passes Milestone Level." *The Guardian* (2013).

Ceballos, Gerardo et al. "Accelerated Modern Human – Induced Species Losses: Entering the Sixth Mass Extinction." *Science Advances* 1:5 (2015): 1–5.

Coole, Diana and Samantha Frost. *New Materialisms: Ontology, Agency, and Politics*. Durham: Duke Universoty Press, 2010.

Crutzen, Paul J. "Geology of Mankind." *Nature* 415:6867 (2002): 2.

Deleuze, Gilles and Félix Guattari. *Anti-Oedipus: Capitalism and Schizophrenia*. New York: Viking Press, 1977.

———. *A Thousand Plateaus: Capitalism and Schizophrenia*. Minneapolis: University of Minnesota Press, 1987.

Ellis, Erle C. et al. "Anthropogenic Transformation of the Biomes, 1700 to 2000." *Global Ecology and Biogeography* 19:5 (2010): 589–606.

Foucault, Michel. *The Order of Things: An Archaeology of the Human Sciences*. London and New York: Routledge, 2002.

Gourmelon, Gaelle. *Book Global Plastic Production Rises, Recycling Lags*. Washington, DC: Worldwatch Institute, 2015.

Hamilton, Clive. "Getting the Anthropocene So Wrong." *The Anthropocene Review* 2:2 (2015): 102–7.

Haraway, Donna. *The Companion Species Manifesto: Dogs, People, and Significant Otherness*. Chicago: Prickly Paradigm Press, 2003.

Hunt, James. "Introductory Address on the Study of Anthropology." *The Anthropological Review* 1 (1863): 1–3.

Kolbert, Elizabeth. "Enter the Anthropocene – Age of Man." *National Geographic* 219:3 (2011): 60.

———. *The Sixth Extinction: An Unnatural History*. New York: Henry Holt, 2014.

Kraeling, Carl Hermann. *Anthropos and Son of Man*. New York: Columbia University Press, 1927.

Latour, Bruno. *Politics of Nature: How to Bring the Sciences into Democracy*. Cambridge, MA: Harvard University Press, 2004.

Law, John. "After Ant: Complexity, Naming and Topology." *The Sociological Review* 46 (1998): 1–14.

Morton, Timothy. *Ecology without Nature: Rethinking Environmental Aesthetics*. Cambridge, MA: Harvard University Press, 2007.

Purdy, Jedidiah. *After Nature*. Cambridge: Harvard University Press, 2015.

Ruddiman, William F. "The Early Anthropogenic Hypothesis: Challenges and Responses." *Reviews of Geophysics* 45:4 (2007): 1–37.

———. "The Anthropocene." *Annual Review of Earth and Planetary Sciences* 41:1 (2013): 45–68.

Steffen, Will, Regina Angelina Sanderson, Peter D. Tyson, Jill Jager, Pamela Matson, Berrien Moore III, Frank Oldfield, Katherine Richardson, Hans Joachim Schellnhuber, and Billie L. Turner. *Global Change and the Earth System: A Planet under Pressure.* Berlin: Springer, 2004.

Steffen, Will, Paul J. Crutzen, and John R. McNeill. "The Anthropocene: Are Humans Now Overwhelming the Great Forces of Nature." *AMBIO: A Journal of the Human Environment* 36:8 (2007): 614.

Turnbaugh, Peter J. et al. "The Human Microbiome Project." *Nature* 449:7164 (2007): 804–810.

Waters, Colin N. et al. "A Stratigraphical Basis for the Anthropocene?" *Geological Society, London, Special Publications* 395:1 (2014): 1–21.

———. "Can Nuclear Weapons Fallout Mark the Beginning of the Anthropocene Epoch?" *Bulletin of the Atomic Scientists* 71:3 (2015): 46–57.

Wolfe, Cary. *What Is Posthumanism?* Minneapolis: University of Minnesota Press, 2010.

Zalasiewicz, Jan, Mark Williams, Will Steffen, and Paul Crutzen. "The New World of the Anthropocene." *Environmental Science & Technology* 144:7 (2010): 2228–31.

Zalasiewicz, Jan et al. "The Technofossil Record of Humans." *The Anthropocene Review* 1:1 (2014): 34–43.

———. "When Did the Anthropocene Begin? A Mid-Twentieth Century Boundary Level Is Stratigraphically Optimal." *Quaternary International* 383 (2015): 196–203.

Part I

NONHUMAN AGENCY

Chapter 1

Carbon Bonds

Coal Economics and Aesthetics

Sean Cubitt

According to Epstein et al., the unaccounted cost of coal – what economists call the externalities associated with mining, transporting and burning it – amount to something between US$ 350 and 500 billion a year in the United States alone, a cost that would, if fully paid for, triple the cost of what is still the largest source of electrical supply in the contemporary world, making it far more expensive than the cost of renewables.[1] And yet coal provides more than 40 percent of the world's electricity.[2] SourceWatch lists twenty-seven forms of pollution from coal, ranging from floods caused by mining activities to air pollution.[3] This coal is ubiquitous in digital media, electronic media more generally, and the mechanical media that preceded them in the form of the energy required to make and run them, and in the form of plastics and other chemical products required to make them. It would be hard to imagine Epstein and his colleagues or SourceWatch being able to conduct the very large-scale research they do, or being able to communicate it to others today, without using those media. We are locked into coal, even in the era of peak oil, and perhaps even more so as a result of the fears of oil scarcity in the near future. In what follows, and in the footsteps of YoHa's remarkable artwork Coal Fired Computers,[4] I would like to present the materiality of coal as an integral part of the history and actuality of media, to suggest some ways that the coal industry and its offshoots prefigured the contemporary reign of finance capital, and to indicate how and why these stories can help us reconsider how we construct and conduct politics and the communications that underpin them in the twenty-first century.

NEW KING COAL

In the reign of Queen Elizabeth the First of England, priors of coastal Northumberland sent coal to London, where the Company of Brewers offered to fire their breweries in the vicinity of Westminster Palace with wood because the Queen "findeth hersealfe greately greved and annoyed with the taste and smoke of the see coles."[5] Not so grieved as all that: within a decade, Good Queen Bess had acquired the landholding of key Tyneside collieries and become the beneficiary of the pollution she so assiduously avoided for herself. Not that anyone given the opportunity would not have avoided it. The "sea-coals" of Newcastle, so called because the seams tumble out onto the North Sea coast, are rich in sulphur and when burnt not only release carbon and particulates but also large amounts of sulphur which, combining with water as sulphuric acid, produces acid rain, which ate away at the clothes and damaged the complexions of her subjects until the Clean Air Act of 1956 finally began to control the industrial as well as domestic use of coal for heating and energy.

In those years, London acquired its nickname of "The Smoke," and its reputation for pea-soup fogs, smog of such deadly thickness that as late as the 1950s the death toll from particularly dense aggregations mounted into the tens of thousands, as was the case in other coal-fired industrial cities like Pittsburgh, and is the case now in Beijing. As Barbara Freese points out, coal's portability and its efficiency compared to wood, made it perfect for steam engines, which in turn made deep mining possible, draining water, operating lifts, and freighting the product to the industrial centres.[6] Cheaper coal meant more railways which meant cheaper coal, in a vicious spiral. Steam-powered bulk shipping of the product of distant colonies fed the imperial factories, and transported the finished goods back to the colonial markets. If the Victorians ever spoke of the "engine of growth," they would undoubtedly have imagined a steam engine powered by coal.

It is not only that transport, which was traditionally included in the sociology and geography of communications well into the 1970s, mediates between distant places; or simply that coal-powered transport enabled a mass market in illustrated magazines and railway station bookstores, or housed vast arrays of advertising hoardings. It is not even that through transport, coal was communicated from its ancient underground home to faraway places, to be burned and redistributed in water and air in a new pattern. It is rather that coal began to make its way into the very fabric of media, from powering the printing presses to providing its colours. Industrial chemistry began, to the extent that such a statement can ever be more than a story indicating a broader social movement, with the 1856 invention of aniline dyes. The nineteen-year-old William Perkin, then following classes in chemistry at the Regents Street Polytechnic, was attempting to synthesize quinine when he accidentally produced a rich mauve dye. Unlike all previous dyes, which came from animals, vegetables of minerals, Perkin's mauve, the first chemically synthesized dye, derived from coal tar, the sticky, toxic residue of the coal gas used to fire street illumination and domestic lighting in the nineteenth century. Perkin made his fortune when the widow of Napoleon III determined to lighten her mourning garb and lit on the new dye as an appropriate step from all-consuming black to a seemly but more colourful afterlife. What the Empress wore today, all Europe wore tomorrow. Synthetic dyes and the fashion industry have a common origin.

There is something magical about the emergence of synthetic colour from the black toxin of coal tar, as Michael Taussig has adumbrated.[7] The rainbow of synthetic hues celebrated by modernist avant gardes in the early twentieth century as much as the blaze and clash of colour in public space saturated with the new paints, printers' inks and dyes set the world alight (and ruined many industries based on the old natural dyes like indigo and madder). Coal's black splintered into as many hues as Newton's prism had extracted from the white light of the sun. At the same time, as Esther Leslie is at pains to demonstrate, the anti-natural move to synthetic colour remained caught in a dialectical relationship with ideologies of nature like the Nazi's "blood and soil," and linked the chemical industry that grew out of Perkin's mauve indissolubly with war, from standardized colours of flags and uniforms to the art of camouflage, in an industry which, while based in colours, rapidly adapted to the manufacture of explosives, many of them used to blast more coal from the earth and to dynamite cuttings for railways to bring more coal to industry.[8]

In 2015, the price of oil plummeted, in part due to deliberate oversupply from Saudi Arabia. The goal was to make North American fracking and extraction of oil from bituminous shale, intended to achieve "energy security," uneconomic and thus preserve the political power of the House of Saud. Coal, which had increased its market share in the wake of the Fukushima nuclear disaster, also saw plummeting prices due to oversupply and dramatic falls in consumption due to the ongoing effects of the 2007 global financial crisis, especially in Germany, in India and most of all in China, which is also concerned to improve its air quality by cutting coal use, and has its own over-producing artisan (effectively unregulated or "wildcat") coal mining sector. In the year of the Paris COP15 conference, which achieved, against all the odds, some formal agreement on constraining climate change through management of fossil fuels, the tumbling price of oil and coal tended in exactly the opposite direction. Given that nine of the ten top companies in the Fortune 500 international listing are in the fossil fuels, automotive and energy businesses, and that spending on coal industry lobbyists was running at US$ 18 million a year between 2008 and 2012[9] (the energy sector as a whole spends US$ 240 million a year in lobbying the US government), it seems likely that economic pressures will trump political sentiment. Since the US Republican Party is the only major nation conservative party to maintain its position of climate change denial[10] and likely to control the House if not the presidency after the next election, while the United Kingdom's Conservative government, during the Paris talks, committed to hundreds of millions of pounds subsidy to diesel-powered electricity generation[11] and support for the fracking industry, that political sentiment seems merely sentimental among governing elites. The coal industry, meanwhile, boasts that carbon sequestration can make coal-fired generators carbon neutral: yet, as Tyfield notes, to date, "not a single full-size coal-fired power plant with CCS [Clean Coal System] is in operation."[12] Indeed, although the Paris COP21 Agreement insisted in Article 2 that CCS would be essential in meeting the goal of restricting global warming,[13] only days before the UK government abandoned manifesto pledges to invest in the technology.[14]

The magic of coal seems very different in this light. In part this may be due to a further aspect of coal, the culture of working-class masculinity and militancy associated with coal throughout the twentieth century. Colliers were the most exploited of workers on both sides of the Atlantic, and in the nineteenth and early twentieth centuries fought back to establish unions based on the personal bonding that they needed to survive life down the pit. If, as Mitchell argues, the rise of oil to prominence was based on ousting miners' unions from their position of political and economic power,[15] the closure of the last UK deep coal mine, also in 2015, indicates the end of a century-long struggle that reached its apogee in Mrs Thatcher's epic war against the miners in the mid-1980s. Coal, Mitchell argues (and in this is echoed by Barbara Freese[16]), is tough to mine and tough to transport. It needs hauliers to move it from pithead to freight wagons, or to load it onto ships. At each point the human factor is likely to organize for its own interests rather than that of the owners. Grinding physical labour, hard-won solidarity, and a tradition of class struggle made the miners the last bastion of an admittedly masculinist and dirty but nonetheless profound political and cultural workplace, for lack of which the young men of ex-mining and manufacturing regions are increasingly disaffected and anti-social.[17] The rise of coal consumption and the demise of organized coal unions are intertwined. This is not the triumph of a new phase of capital overhauling the old industrial mode. We should not feel nostalgia for the dirt or the hard work, the lung diseases, the machismo. The bad new days are only the equal of the bad old days. In the increasingly rapid realization of a long-delayed and deep-seated tendency in capital, this disaggregation of consumption from production is not contingent or marginal: the very energy and plastics that coal made and makes possible form the contemporary electronic media that enable the emergence of capital's intimate truth into the daylight. Financialization is unthinkable without coal.

COAL DERIVATIVES

Carbon monoxide. Inimical to life. By-product of coal combustion, carbon monoxide is used throughout the chemical industry to produce acetic acid, aldehydes, methanol and phosgene, a precursor in plastics production. The disaggregated sulphur and trace metals, notably iron and mercury, are extracted for hundreds of product lines from solvents to lasing media in high-powered lasers, as used in long-distance fibre-optic backbones of the internet. Coal is even a major source of the hydrogen used in fuel cells. SourceWatch notes, among heavy metals found in coal seams, lead, mercury, nickel, tin, cadmium, antimony, and arsenic, uranium and isotopes of thorium and strontium, all of which are concentrated in the fly ash particulate matter released when the coal is burnt.[18] Much of the ash is captured in scrubbing towers in larger power plants, and many of the metals are extracted, but much of the remaining sludge is uneconomic to cleanse and finds its way, through circuits of specialist recovery and cleansing plants, into landfill. The centuries-long history of coal-burning has left us a legacy of toxic metals as well as plastic waste, acid rain and methane emissions from which the planet struggles to recover. To the extent that the internet is a planetary phenomenon, any of our messages may be handled by coal-powered servers and routers, even if no coal is burnt locally. As the atmosphere is no respecter of borders, plumes of smoke can cross oceans. Where pollutants reach the water cycle, as in the case of sulphur dioxide from bituminous coal, the effects are global. The planet-spanning scale of telecommunications is matched by the by-products of the power used to run them and the plastics employed to protect them from the environment they have helped to produce. From the smoke that "greved" the Queen at Westminster to the buried waste of power stations waiting to leach into rivers and oceans, coal is not an inert commodity, nor a raw material for communication: it communicates, from ancient swamps to future cataclysms, in and with the natural, human and technological worlds.

At which point it is important to consider the kinds of traffic carried by the internet and telecommunications more generally. According to Cisco, "By 2016, global IP traffic will reach 1.1 zettabytes per year, or 88.4 exabytes (nearly one billion gigabytes) per month, and by 2019, global IP traffic will reach 2.0 zettabytes per year, or 168 exabytes per month."[19] While we might blame high-bandwidth consumer services like streaming video and online gaming for the massive increases, in 2013, one estimate had it that that only 38.5 percent of internet traffic originated with humans: the rest were bots, among which cookie traffic from advertisers and automated trading bots from the world's finance markets formed a major slice, along with smaller numbers of surveillance bots from governmental sources.[20] In 2015, humans made up a larger proportion, just over half of web traffic, but in the context of overall skyrocketing growth in use this only indicates that bot traffic is also growing, and fast.[21] Alongside business-to-consumer and business-to-business communications, including delivery of electronic consumer goods like videos and ebooks, the trade in futures and other derivatives is a real-time market which works on small margins and rapid changes in market conditions for particular commodities and bundles of commodities. This is the kind of live spot market whose emergent chaotic systems failure helped collapse the entire eastern seaboard power supply in Canada and the United States in 2003 and the May 2010 Flash Crash which wiped trillions of dollars off Wall Street in slightly over half an hour. Coal futures are part and parcel of this market, 40 percent of whose energy needs it supplies.

The volatility of derivatives markets and their centrality to the global financial crisis – whose eight-year duration bends the meaning of the word crisis beyond breaking point – are often held to mark a new era in capitalism, tied to the broad analysis of neo-liberalism, the economic dogma embraced by Thatcher and Reagan in the 1980s and now the consensual faith of national governments and international economic and governance bodies worldwide. There is strong evidence, marshalled by David Harvey, that neo-liberalism is not in fact an economic practice at all but a political one, aimed at ending the long struggle, from the fall of the European crowned heads in 1914 to the breakdown of state socialism in the Soviet bloc in 1989, to create and maintain a welfare state and other basic rights for the working class.[22] For Harvey and others, this marks the emergence of finance capital as the next stage of capital after the industrial era and the Fordist-Keynesian mid-twentieth-century age of consumerism. For others, including Maurizzio Lazzarato, financialization represents only the re-emergence of capital's true orientation, which is to produce money from money.[23] Capital's goal is profit. Industrialization provided one way to make an initial investment grow. Trapped by the falling rate of profit, Keynesianism provided a new way to ensure that increase in consumption which would keep investment in the profit side of the ledger. Manufacture and consumption were merely means to secure the central goal, making more money out of money.

Coal's role here is important historically as well as in its current role as engine of the internet. The native necessity of capital is growth. Marx explains this as being required by the falling rate of profit. As technologies develop, displacing workers, they produce increasing numbers of commodities, but at a unit cost which shrinks on the well-known principle of "pile 'em high and sell 'em cheap." Because the amount of profit on each sale diminishes, to secure growth capital needs to sell more units. To do so, it has to produce more. But since in the process it has displaced the industrial workers whose wages used to pay for the commodities they produced, capital tends to crises of overproduction. To solve this problem and secure the growth without which its crisis becomes terminal, capital offers loans so that the poor can afford to buy their products. But a loan is only a pledge to pay today with money borrowed against future wages, which the logic of technological improvement suggests is a bad bet: fewer people will have well-paid work in future, and less disposable income due to the privatization of the old welfare system. An economy based on growing household debt, especially among those who can never hope to repay it, was a central cause of the mortgage bubble that instigated the 2007 crash, and is being rebuilt again as neoliberal finance ministers globally use debt once again as a replacement for the missing growth that capital can no longer guarantee.[24] The objective of speculators in the derivatives market, in the words of the Coal Trading Association, "is to make a profit from taking on the risk of price fluctuation that commercial users do not want."[25] Since future rates of profit cannot be known in advance, the risk associated with future prices can be sold on as a separate product, giving speculators a tranche of future profits, but also enabling them to sell on the risk. Speculation in debt repayment works in exactly the same way: to create profit today from the trade in repayments tomorrow, including the option to make a profit from a falling market price by selling on before the loss becomes apparent. Extracting value is the goal: the means may change, but the goal remains the same.

There are then both continuities and discontinuities in the new capitalism. Rosa Luxemburg reanalysed the period of primitive accumulation that Marx had addressed in the formation of capital in the early modern period when common land was taken from the peasants, privatized and used to extract wealth.[26] Luxemburg's insight was that this process continued in the equally brutal dispossession of colonized peoples in the age of the European empires. In our time this process continues with the privatization of common goods and services, especially those created by workers' struggles for the welfare state, but also in the creative legislation that encourages the exploitation of mineral and fossil resources. Massive government subsidies for both extraction industries and financial services, paid for by selling off common goods like health, security and prison services, are ultimately paid by seizure of common assets, including the naturally occurring geology, water and air, and the permissions granted to capital to pollute these resources at no cost to itself, paid for again in the ill-health and impoverishment of the increasingly disenfranchized poor. Accumulation by dispossession, to use Harvey's phrase, underwrites two other principles of contemporary capital, one of which was also already noted by Marx under the rubric of "extended reproduction," by which he meant the investment of profits into growth rather than into meeting the needs of the present.[27] In the industrial phase of capital, that made sense: the present in effect lent money to the future in return for the promise of growth. Today however the massive profits extracted by financial and other corporations are increasingly removed from the reinvestment cycle, precisely because the use of debt as surrogate for growth means that interest rates have been at zero for years, meaning that the return for investing in the future is now nil. As a result huge amounts of money, estimated in the trillions of dollars, are sequestered in offshore financial havens, doing nothing, even as the world enters a third round of the current crisis with massive food shortages – associated with climate change – in Asia and Africa, and major wars raging across central and North Africa and the Middle East, again associated with climate change, drought and the brutal race to extract mineral wealth without benefit to the peasant natives of Congo, Angola and other war-torn nations. We have learned not to expect generosity from those who extract this wealth. It is quite possible that corporations, which benefit from the legal status of persons in intellectual property and contract law, have in fact sacrificed their right to be considered human in their rabid and suicidal race to accumulate wealth they do not intend to use at the expense of the commons and common decency. Privatization is synonymous with privation.

The commodity futures market now governs the coal industry. The trade in derivatives provides the coal industry, like other extraction industries, with the investment capital it needs to maintain growth. Since capital is absolved of consideration for the common good, that investment is premised on extracting more fossil fuel, and ensuring a market for it, rather than, for example, mitigating global warming. Futures prices, as measure of the likely market for coal, indicates the level of investment the market is prepared to put into the industry, thus effectively overriding any public policy moves by the momentum of the demand for profit. The common good we might aspire to derive from the common heritage of fossil coal is thus removed from the social in order to become private profit. This privatization – and privation – is the more shocking since these profits are, under the current condition of austerity capital, consistently removed from circulation, rather than reinvested as in earlier epochs of industrial and service capital, and retained outwith the general economy, even if that risks the very principle of growth on which capital depends in order both to survive and to legitimate itself. To privatize and then sequester profit is inhuman to the extent that it is suicidal: not only endangering human lives, but the very financial machinery on which it depends: no human creature would act in such a self-destructive way.

It also indicates a strange form of temporality through which capital extends its rule over history. Coal, in common with almost all our energy sources, is the product of sunlight. Millions of years ago, vast forests sank into swamps, were covered over, and in the dark, heat and pressure became rich seams of black carbon-rich rock. The coal we burn today is the gift of ancient sunlight. If there is debt involved, then we owe a debt to that ancestral forest. It is in this light that we can understand the otherwise baffling belief of Chip Beeker, a leading Alabama Republican debating proposed Environmental Protection Agency regulation of coal-fired power plants, that "coal was created in Alabama by God, and the federal government should not enact policy that runs counter to God's plan." While Twitter lit up with observations that God might have placed fossils in the ground so we could learn about evolution, few noted the true paradox of Beeker's faith: that coal beds (and oil fields) represent the most successful carbon sequestration project, without which Earth would have become uninhabitable long ago. If God had wanted us to burn coal, he wouldn't have buried it. And yet there is a grain of truth in Beeker's belief: coal is a gift from the universe, and should therefore be treated with respect. The immense number of chemicals already derived from coal indicate that we have scarcely begun tapping our chemical heritage; and that by burning it in vast quantities, we are destroying both the legacy of the past and the possibilities of the future. In turn, this indicates the temporal structure of capital lurking under the engines of perpetual technological and social change and the principle of growth, which would seem to make capital and the faith in progress an entirely future-oriented process. Instead, in every blazing lump of coal, we witness the sacrifice of both past and future to the present. Capital turns the obligations we owe to our ancestors and our offspring as natural human beings into debts owed by the past and the future to the present. The present, in this guise is a single point moving through time, a point which in itself remains unchanged. We might choose to see in this conception of time the diagram for the mechanical moving image of the analogue era, while the flying scan of digital imaging would represent only its acceleration.

CARBON AESTHETICS

During the Massey conference on Working with Nature in April 2015, Australian eco-photographer Chris Orchard discussed Freud's narrative of the infant's forced separation from primary union with mother's body and world. That first trauma, Orchard suggested, is re-enacted in every photograph, ripping the instant from its home in the wholeness of flux. The trauma of the instant forcibly isolated from the flow of time is the tragedy of the still image. We might add that it is too the basis of its radical incompletion. The still photograph presents itself as if whole and entire, but is always divorced from how it became, and what it might become: it is purely actual, without potential, but in precisely that case it acquires a potential that the actual moment seized in the photo did not have: the chance to enter into dialogue with other, alien moments in time whenever someone looks at it. The moving image, we might continue, appears in the photographic laboratories of the last decade of the nineteenth century in order to heal both the trauma of separation and the radical new potentiality of the stilled image. This is the dialectic instigated by capital's seizure of the perpetual present as its site of accumulation: dependent on investment into futures to resolve the risks of the present, yet compulsively stripping profit from the future in order to accumulate now.

This fundamentally unstable temporality is mirrored first in the still image, then in the attempt to replace each still with a successor in an endless movie; and finally in the ubiquitous imaging of twenty-first-century digital media with their scans determining that even the individual frame can no longer pretend to wholeness. Multiplication and incompletion underpin capital as it underpins its image regimes.

Coal's role in this is momentarily obscured by the manufacturing processes of the substrates of cinema. Nitrate film stock, the notorious early film capable of spontaneous combustion, was a close relative of nitroglycerin, the explosive. Its chemical antecedents include a number of chemicals derived from coal and coal gas. The colours and dyes applied throughout the colour era, from hand-painted strips in the first years of the twentieth century to Technicolor in mid-century and on to contemporary colour filters in digital projectors all derive from Perkin's initial discoveries. Coal-derived plastics coat and protect wiring and cabling, lamp and camera housings, lens coatings and internet routers, and provide many of the elements in new ceramic- and glass-based digital components. One of the most condensed forms of carbon, coal's complex chemistry derived from aeons of growth, decay and geological compaction is disassembled into its components and rewired into a global communications network of immense complexity, among whose most important roles is to provide the infrastructure for investment trades.

Coal thus oscillates between two distinct modes of existence. On the one hand, it is a mineral with certain chemical qualities, which have made it historically a vital component in the building of the modern world, from railways to power generation and the formation of the chemical industry. At the same time it functions as a commodity, exchangeable for any other, equal to any other, and to that extent without any specific qualities at all save its price. We recognize that the first aspect, the use value of coal, is already multiple and temporal, and that there is no simple identity that can be ascribed to its multiple relocations and uses. Perversely the exchange value of coal does ascribe such an identity, attempting to anchor coal to its present – for example by pricing up reserves in hand – and ignoring both its use and abuse. External to both accounts are the immense costs of mining, burning and chemical manipulations of coal, but perhaps most of all the claim of coal itself to a stake in its own uses and exchange.

CONCLUSION

It would be absurd to suggest that coal mines, coal deposits, slag heaps and soot should have a stake in public life; that they should vote or address assemblies. Yet we have already welcomed one nonhuman agent into the political life of nations, and given it immense authority in rule, to the extent that we say to one another that we have no alternative but to obey the dictates of the Market, that immense network of computers into which we plug interchangeable bank employees, from clerks to CEOs. The Market cyborg rules. This explains the pursuit of profit even to the point at which it risks the survival of the human race. But if we have given the Market such an immense say in our human affairs, why can we not give an equal, and greater voice, to the nonhuman environment? One answer is Wittgenstein's: "If a lion could speak, we could not understand him."[28] The fact is however that lion's – and climatic systems, glaciers, forests, polar bears and cockroaches – can and do speak, but that we have no desire to listen when they warn us of what is happening to them and to us.

The problem then is one of communication. We have surrendered the greater part of our communications media to advertising and trade. Very little of a communications common is left, and much of that is taken up, understandably if disturbingly, by political and religious factions driven by desperation and exclusion, and by the cries for help of the isolated and stressed created by the destruction of the common good. Inventing new means of communication with the natural world will seem to these as a trivial pursuit compared to the immediate danger of their own concerns. Yet that is precisely why communication is the central platform of a new politics. Since public administration has failed, alongside capitalist economies, we must work on the infrastructure on which both depend, the media. Communication is the goal but also the means of a new, open polity, one in which coal will take its place not as exploited reserve but as active participant. It is in our gift, a gift which derives from the debt we owe both past and future, to honour and recognize the minerality in the materiality of media, and in that recognition to allow coal and the other materials to speak their part, to make the materiality of media the basis for an environmental and communicative politics for the twenty-first century.

NOTES

1. Paul R. Epstein, Jonathan J. Buonocore, Kevin Eckerle, Micahel Hendryx, Benjamin M. Stout III, Richard Heinberg, Richard W. Clapp, Beverly May, Nancy L. Reinhart, Melissa M. Aherne, Samir K. Doshi, and Leslie Glustrom, "Full Cost Accounting for the Life Cycle of Coal," in *Annals of the New York Academy of Sciences*, Vol. 1219 (New York: New York Academy of Sciences, 2011), 73–98.
2. Greenpeace, *The True Cost of Coal* (Amsterdam: Greenpeace, 2011).
3. SourceWatch, "Environmental Impacts of Coal" (Madison: Centre for Media and Democracy, 2015), http://www.sourcewatch.org/index.php/Environmental_impacts_of_coal.

4. YoHa, *Coal Fired Computers* (2010), http://yoha.co.uk/cfc.

5. Robert L. Galloway, *A History of Coal Mining in Great Britain* (London: Macmillan, 1889), 24.

6. Barbara Freese, *Coal: A Human History* (New York: Random House, 2003).

7. Michael Taussig, "What Color Is the Sacred?" *Critical Inquiry* 39 (2006): 28–51.

8. Esther Leslie, *Synthetic Worlds: Nature, Art and the Chemical Industry* (London: Reaktion Books, 2005).

9. OpenSecrets, "Coal Mining" (Washington, DC: Center for Responsive Politics, 2015), https://www.opensecrets.org/lobby/induscode.php?id=E1210.

10. Sondre Båtstrand, "More than Markets: A Comparative Study of Nine Conservative Parties on Climate Change," *Politics and Policy* 43:4 (2015): 538–61.

11. Kiran Stacey, "UK Turns to Diesel to Meet Power Supply Crunch," *Financial Times*, 3 November 2015, http://www.ft.com/cms/s/0/0f664c78-821b-11e5-8095-ed1a37d1e096.html.

12. David Tyfield, "'King Coal is Dead! Long Live the King!': The Paradoxes of Coal's Resurgence in the Emergence of Global Low-Carbon Societies," *Theory Culture and Society* 31:5 (2014): 59–81.

13. Global CCS Institute, "Round-Up on Negotiations for Saturday, 12 December 2015," *Global CCS Institute* (Melbourne: Global CCS Institute, 2015), https://www.globalccsinstitute.com/news/institute-updates/latest-news-institute-cop21.

14. Damian Carrington, "UK Cancels Pioneering £1bn Carbon Capture and Storage Competition," *The Guardian*, 25 November 2015, http://www.theguardian.com/environment/2015/nov/25/uk-cancels-pioneering-1bn-carbon-capture-and-storage-competition.

15. Timothy Mitchell, *Carbon Democracy: Political Power in the Age of Oil* (New York: Verso, 2011).

16. Freese, *Coal*.

17. Mike Savage, *Social Class in the 21st Century* (London: Penguin, 2015).

18. SourceWatch, "Environmental Impacts."

19. Cisco, "Cisco Visual Networking Index: Forecast and Methodology, 2014–19 White Paper" (San Jose: Cisco Systems, 2015), http://www.cisco.com/c/en/us/solutions/collateral/service-provider/ip-ngn-ip-next-generation-network/white_paper_c11-481360.html.

20. Meaghan Neal, "Only 38 Percent of Internet Traffic Is from Real Humans Now," *Motherboard*, 12 December 2013, https://motherboard.vice.com/blog/only-38-percent-of-internet-traffic-is-from-real-humans-now.

21. Igal Zeifman, "2015 Bot Traffic Report: Humans Take Back the Web, Bad Bots Not Giving Any Ground," *Incapsula*, 9 December 2015, https://www.incapsula.com/blog/bot-traffic-report-2015.html.

22. David Harvey, *A Brief History of Neoliberalism* (Oxford: Oxford University Press, 2005); David Harvey, *The Enigma of Capital and the Crises of Capitalism* (Oxford: Oxford University Press, 2010).

23. Maurizio Lazzarato, *Governing by Debt*, trans. Joshua David Jordan (New York: Semiotxt(e), 2015).

24. Press Association, "Economy Concerns as Household Debt Rises to £40bn in Latest Figures," *The Guardian*, 22 December 2015, http://www.theguardian.com/uk-news/2015/dec/22/economy-concerns-as-household-debt-rises-to-40bn-in-latest-figures.

25. Coal Trading Association, "Coal Market – Frequently Asked Questions" (Washington, DC: Coal Trading Association, 2010), https://coaltrade.org/documents/CTACoalMarketFAQ3-10.pdf.

26. Rosa Luxemburg, 1913, *The Accumulation of Capital*, trans. Agnes Schwarzschild (London: Routledge, 1951).

27. Harvey, *Brief History*.

28. Ludwig Wittgenstein, *Philosophical Investigations*, trans. G. E. M. Anscombe, P. M. S. Hacker and Joachim Schulte, 4th ed. (Oxford: Basil Blackwell, 2009), 120.

BIBLIOGRAPHY

Båtstrand, Sondre. "More than Markets: A Comparative Study of Nine Conservative Parties on Climate Change." *Politics and Policy* 43:4 (2015): 538–61.

Carrington, Damian. "UK Cancels Pioneering £1bn Carbon Capture and Storage Competition." *The Guardian*. 25 November 2015. http://www.theguardian.com/environment/2015/nov/25/uk-cancels-pioneering-1bn-carbon-capture-and-storage-competition.

Cisco. "Cisco Visual Networking Index: Forecast and Methodology, 2014–19 White Paper." San Jose: Cisco Systems, 2015. http://www.cisco.com/c/en/us/solutions/collateral/service-provider/ip-ngn-ip-next-generation-network/white_paper_c11-481360.html.

Coal Trading Association. "Coal Market – Frequently Asked Questions." Washington, DC: Coal Trading Association, 2010. https://coaltrade.org/documents/CTACoalMarketFAQ3-10.pdf.

Epstein, Paul R., Jonathan J. Buonocore, Kevin Eckerle, Micahel Hendryx, Benjamin M. Stout III, Richard Heinberg, Richard W. Clapp, Beverly May, Nancy L. Reinhart, Melissa M. Aherne, Samir K. Doshi, and Leslie Glustrom. "Full Cost Accounting for the Life Cycle of Coal." *Annals of the New York Academy of Sciences* 1219 (2011): 73–98.

Freese, Barbara. *Coal: A Human History*. New York: Random House, 2003.

Galloway, Robert L. *A History of Coal Mining in Great Britain*. London: Macmillan, 1889.

Global CCS Institute. "Round-Up on Negotiations for Saturday, 12 December 2015." *Global CCS Institute*. Melbourne: Global CCS Institute, 2015. https://www.globalccsinstitute.com/news/institute-updates/latest-news-institute-cop21.

Greenpeace. *The True Cost of Coal*. Amsterdam: Greenpeace, 2011.

Harvey, David. *A Brief History of Neoliberalism*. Oxford: Oxford University Press, 2005.

———. *The Enigma of Capital and the Crises of Capitalism*. Oxford: Oxford University Press, 2010.

Lazzarato, Maurizio. *Governing by Debt*, trans. Joshua David Jordan. New York: Semiotxt(e), 2015.

Leslie, Esther. *Synthetic Worlds: Nature, Art and the Chemical Industry*. London: Reaktion Books, 2005.

Luxemburg, Rosa. *The Accumulation of Capital*, trans. Agnes Schwarzschild. London: Routledge, [1913] 1951.

Mitchell, Timothy. *Carbon Democracy: Political Power in the Age of Oil*. New York: Verso, 2011.

Neal, Meaghan. "Only 38 Percent of Internet Traffic Is from Real Humans Now." *Motherboard*, 12 December 2013. https://motherboard.vice.com/blog/only-38-percent-of-internet-traffic-is-from-real-humans-now.

OpenSecrets. "Coal Mining." Washington, DC: Center for Responsive Politics, 2015. https://www.opensecrets.org/lobby/induscode.php?id=E1210.

Press Association. "Economy Concerns as Household Debt Rises to £40bn in Latest Figures." *The Guardian*, 22 December 2015. http://www.theguardian.com/uk-news/2015/dec/22/economy-concerns-as-household-debt-rises-to-40bn-in-latest-figures.

Savage, Mike. *Social Class in the 21st Century*. London: Penguin, 2015.

SourceWatch. "Environmental Impacts of Coal." Madison: Centre for Media and Democracy, 2015. http://www.sourcewatch.org/index.php/Environmental_impacts_of_coal.

Stacey, Kiran. "UK Turns to Diesel to Meet Power Supply Crunch." *Financial Times*, 3 November 2015. http://www.ft.com/cms/s/0/0f664c78-821b-11e5-8095-ed1a37d1e096.html.

Taussig, Michael. "What Color Is the Sacred?" *Critical Inquiry* 39 (2006): 28–51.

Tyfield, David. "'King Coal is Dead! Long Live the King!': The Paradoxes of Coal's Resurgence in the Emergence of Global Low-Carbon Societies." *Theory Culture and Society* 31:5 (2014): 59–81.

Wittgenstein, Ludwig. *Philosophical Investigations*, trans. G. E. M. Anscombe, P. M. S. Hacker and Joachim Schulte. 4th ed. Oxford: Basil Blackwell, 2009.

YoHa, *Coal Fired Computers* (2010). http://yoha.co.uk/cfc.

Zeifman, Igal. "2015 Bot Traffic Report: Humans Take Back the Web, Bad Bots Not Giving Any Ground." *Incapsula*, 9 December 2015. https://www.incapsula.com/blog/bot-traffic-report-2015.html.

Chapter 2

Consider the Lawnmower

Aesthetics, Politics and Entanglements of Suburban Nature

Nicholas Holm

Although our dreams of the nonhuman are often evocative of the Amazon, Atlantic or Antarctica, in practice our everyday entanglements with "nature" take place in much more mundane surroundings and on much more modest scales. Consequently, although ecological interest in the Anthropocene may often be focused on "big picture" tragedies – deforestation, extinction, ocean acidification, ozone depletion – for most inhabitants of the rich countries of the "developed" world, our direct interaction with the living nonhuman world manifests in places such as the backyard, rather than the tundra or the rainforest. Given the centrality of the backyard to our directly lived and experienced ecological entanglements with this living nonhuman world, suburban spaces need to be understood as a central site at which entanglements of humans and living nonhumans are played out in the Anthropocene. In such a context, apparently mundane practices, such as lawn mowing, appear as crucial ethical, aesthetic and political interventions. By approaching lawn mowing as an engagement with local manifestations of the vibrant, agential nonhuman world – one by which many of us make and remake our connections with our most immediate "natural" environment – we can not only interrogate the well-mapped disjuncture between our idealistic visions and lived material practices, but also envision how we might begin to take responsibility for the possibilities of human agency in the Anthropocene: an era where human influence has become an inescapable fact.

ENCOUNTERING THE LOCAL NONHUMAN WITH A LAWNMOWER

Although when considered in light of larger ecological concerns it may seem an exercise in overstatement, Lesley Head and Pat Muir's declaration that "lawns are controversial" certainly captures some of the quiet passion that can unexpectedly accrue to questions of yard maintenance.[1] Historically, the lawn has been understood as alternately a form of human domination or investment in the nonhuman world, and specifically the living nonhuman world (for the purposes of this essay, the term "nonhuman" will refer specifically to the living aspects of the nonhuman). Existing at the intersection of multiple forces and scales, the lawn is formed out of a confluence of social, aesthetic, economic and nonhuman factors all of which carry the potential for disagreement. In this sense, the lawn is even more complicated than the "techno-social system" proposed by Paul Robbins and Julie Sharp, because it is produced not only by the convergence of technologies and social norms, but also out of the unpredictable interruptions of the nonhuman world.[2] Depending upon which aspects of this calculus are regarded as desirable and which exert the most influence in a given context, a lawn can be the scene of ecological encounter, an environmental black hole of chemicals and irrigation, a marker of both community solidarity and conformist oppression, the violent domination of nature by man or any combination of the above. As a consequence of these multiple orientations towards the lawn, debates can take on a level of intensity that seem all out of keeping with the possible politics and consequences of maintaining a parcel of turf grass.

Given these conflicting perspectives, lawn mowing thus provides a case study by which to imagine both more and less desirable forms of ecological entanglement in the Anthropocene. Approaching lawn mowing as a form of (incredibly) small-scale terraforming, we can extrapolate out from the mundane context of yard work to larger questions of how human agency might shape the world around us. In particular, lawn mowing models an engagement that has been afforded some importance to questions of aesthetics in the Anthropocene. More pessimistic accounts of lawn mowing present it as a process of relentless homogenization exemplified by the endless, featureless suburban lawnscape, while more affirmative accounts attempt to account for the possibilities of diversity and difference in the context of the lawn. Such debates provide multiple models of human–nonhuman interaction where human agency can be understood as a force that works in conjunction with the nonhuman to shape a shared world in ways that go beyond simple exploitation. Consequently, lawn mowing can provide a productive means for thinking through the responsibilities and repercussions of human–nonhuman entanglements in the Anthropocene – an era when human agency is always already a fact – because of its conceptualization of human agency in terms beyond destruction.

The key means by which human agents intervene and engage with the nonhuman in the lawn space is through the vector of the lawnmower. While other inputs are certainly important in the upkeep of the lawn in many contexts – in particular the application of a range of industrial chemicals driven by the lawn-chemical economy[3] – it is the lawnmower that is both absolutely essential to the form of the lawn and almost completely inappropriate in any other setting. Hence, while irrigation, fertilizers and herbicides may be used in a number of botanical contexts,[4] the lawnmower only makes sense in the singular situation of the lawn. It is a one-use tool and that use is transformation of grass into lawn through the application of rapidly rotating blades. Indeed, so tightly entwined are the lawnmower and the modern lawn, that the latter is almost unthinkable without the former. In a historical sense, "the hand-operated rotary lawnmower made small lawns possible" and thereby created the conditions by which suburban homeowners could replicate environments previously only possible through the retention of a staff of gardeners and grounds staff.[5] The lawnmower, we could say then, is the instrument by which the lawn as mass suburban phenomena – the conditions for the current discussion – is made possible and it is through the ongoing use of the lawnmower that human actors continue to reconstruct the lawn in aesthetically, socially and ecologically entangled ways.

My own lawn mowing practice is shaped by my stubborn insistence on continuing to use a hand (or push) mower which uses a geared reel to transfer part of my physical effort into the circular motion of the blade cylinder. The critical perspective that informs this chapter is no doubt informed by my choice. As opposed to the dominant gasoline-powered models, hand mowers do not consume fossil fuels or produce air pollution (they are also somewhat quieter than externally powered mowers, though I can confirm that the peaceful solace that some seem to attribute to hand mowers is, to a substantial extent, the product of nostalgic distance from the actual practice). However, while I am more than willing to claim environmental one-upmanship as a side benefit, I should admit that ecological consciousness is not my prime motivation. Rather, my obstinate commitment to the hand mower is largely conditioned by both an awareness of a need for physical exercise and the unclear notion that the best way to spectate cricket is over the transistor radio while pushing a lawnmower. My engagement with my lawn is thus informed by a worryingly retrograde cultural vision populated by obsolete technology and enabled by my location in provincial Aotearoa New Zealand where not only is a hand mower still almost socially acceptable, but where the price of a lawn is not yet out of reach on a university salary. Both my oddly specific cultural motivation and the larger economic context are indicative of how lawn mowing is always caught up in wider structures and flows. Nor do I wish to valorize the use of a hand mower over other models, which would reproduce a worryingly ableist framework: my engagement with this older technology is permitted by my particular embodiment. Moreover, while the use of externally powered lawn maintenance instruments implicates lawn practices in larger discussions of neighbourhood etiquette, energy use and climate change, such modes of engagement nonetheless retain the attentive interaction with the local nonhuman environment that informs my argument.

Lawn mowing is thus never just an individual or pragmatic act: it is a deeply social and aesthetic task caught up in much wider contexts. By this I mean when we mow lawns, we do so as part of a community and in ways that involve more than just the immediate task at hand. Lawn mowing involves not simply trimming turf grass but doing so in a way that reflects both the deeply felt influence of the surrounding society and the potential for personal expression and pleasure. The lawnmower is therefore not simply a means of landscaping. It is also as means of expression via the medium of "the lawn as the common people's art form"[6]: an articulation of the lawn that recognizes its socially shared and aesthetic aspects. In these social terms, the lawn is often experienced in terms of an obligation to the community as part of what Robbins and Sharp refer to as the "moral economy of lawns"[7]: the constellation of duties and judgements that come to bear on the lawn space. An individual human agent does not simply decide how they would like to manage their lawn, particularly in the case of shared suburban frontage, but is instead subject to "neighbourhood peer pressure and social conformity" that predominantly favours uninterrupted, evenly cut turf grass.[8] As a consequence, the lawn frequently operates as an ecological arrangement "that is under constant collective scrutiny and carries great moral weight" that can fall upon those who flout their lawn duties.[9] For many, this sense of social pressure means lawn mowing is experienced as an unwanted obligation to the community, and a manifestation of, in the words of American historian, Albert Way, "all that is wrong with suburbia – a mindless conformity that smothers both individuality and community, bringing the American public to its knees at the altar of commerce and consumerism."[10] If lawn mowing is the common people's art form, then for many it is bad art.

LAWN CRITIQUE: SUBURBIA, CONFORMITY AND MASS NATURE

This criticism of the lawn as the archetypal expression of blandness and homogeneity is a cultural manifestation of "lawn critique," wherein the lawn is singled out as a symbol of one of the most derided of suburban qualities: conformity. Connected historically and culturally to the suburb,[11] the lawn comes to represent the failure of the suburbs. Thus, in contrast to the idea of the garden, which is often partially redeemed by its status as a historical site of transcendence, the potential for internal variation, or an ethical interaction of human and nonhuman participants,[12] the lawn is figured in cultural commentary as an utterly despoiled environment. If the garden is sometimes considered a "cultural [form] designed to shape and contain nature"[13] or the expression of human power and exert control over the nonhuman world, then the lawn is something much worse. For example, in comparison to the relatively benign vision of the garden, Robert Riley argues that the lawn constitutes the most extreme state of human domination over the nonhuman:

> The lawn is a symbol of the ultimate taming of nature and human behaviour. We can see the progressive transformation from forest to forest glade, to meadow, to garden, to lawn as a metaphor of increasing control over, or sublimation of, the raw sexual content of nature. In this progression, the garden is middle ground, where sexuality is controlled but still potent and available

Thus, while it may, in theory, make sense to consider the cultivation of lawns as a specialized form of gardening, discussions of gardening and lawn-maintenance tend to regard the latter as distinct and darker art, where there is a marked intensification of the will to control seen in gardening.[14] Putting to one side "the raw sexual content of nature," what we can see here is that whatever the possible crimes against nature that are found in the garden, the lawn is an even greater offender: an exponential extension of the logic of human control and standardization of nature as part of a "war between man and nature": a masculine struggle where battle is waged with weapons such as herbicides, pesticides and fertilizers.[15]

This critique of the lawn as an icon of suburban conformity is expressed perhaps most virulently in popular nature writer Michael Pollan's 1989 essay, "Why Mow? The Case Against Lawns." While admittedly over two decades old, the continuing resonance of its argument demonstrates the slow-moving nature of lawn politics and the entanglement of ecological, social and aesthetic aspects. In the essay, Pollan recounts his own path towards enlightened opposition to the lawn in heroic terms: lauding his father's stance as a "lawn dissident" against the stultifyingly respectable standards of 1950s Long Island and characterizing his own rejection of the mower as part ecological enlightenment and part aesthetic judgement: "I tired of the endless circuit, pushing the howling mower back and forth across the vast page of my yard, recopying the same green sentences over and over." Drawing on the garden-lawn distinction discussed earlier, Pollan asserts that whereas his garden requires him to learn, engage and adapt to local environmental specifics, the lawn is simply the erasure of local particularity and the brute imposition of conformity. The lawn appears here as not just a denial of "nature," but also an aesthetically impoverished site: it is an expression of "rigid conventions," "anonymous [and] massified," a "vehicle of consensus" rather than "an arena of self-expression," and, perhaps most damningly, "lawns are a form of television." Imagined in such terms, the lawnmower becomes more than just a tool of human domination of the nonhuman; it is also an instrument of aesthetic ruination. Consequently, even when Pollan attempts to shift into a more ecologically minded critique of lawn environments, the aesthetic register continues to shape his condemnation of lawn mowing.

In particular, this narrative of decline is premised on the assumption that human intervention into the nonhuman – specifically collectively determined intervention – leads to degradation. As a waystation between wilderness and lawn, the garden is tolerable as a site of retained difference and specificity. The lawn, however, is damned by its formal homogeneity, in terms strikingly similar to those historically associated with mid-century mass culture critique. In this argument, Pollan replicates both the terminology and logic of a figure like Dwight MacDonald who argues that mass production reduces all aesthetic output to stultifying, pointless mass-produced culture.[16] In parallel fashion, for Pollan, lawn culture reduces local environments to "subdued, homogenised, dominated" mass-produced nature. Moreover, both Pollan and MacDonald explicitly state the problem of the "darker, more coercive side" of democracy, where the victory of majority tastes leads to aesthetic deprivation. Thus, it is not just the human that ruins the nonhuman, here, but the human collective rather than individual. Behind Pollan's denunciation of the suburban lawn lurks a fear of collective intervention as the inevitable cause of the fall from beautiful nature to soulless lawn. In light of such a perspective, if the lawn were to serve as a model for approaching the larger context of the Anthropocene then the consequences would be grim indeed as human intervention leads to the reduction of the nonhuman to a charmless, pointless and ultimately lifeless environment: a form of "mass nature" to mirror the abuses of mass culture.

Understood in this way, lawn mowing appears as a particular form of ecocritic Paul Carter's articulation of colonialism as "ground clearing": lawn environments are after all originally a colonial invention enabled by the expansion of landholding to previously disempowered sections and members of imperial society in colonial contexts, such as the United States, Australia and New Zealand.[17] Following Carter, the lawn is an almost literal manifestation of his account of the "civilised world, as an ideally flat space."[18] The taming of previously unruly nature under the spinning blades of imperial progress; making it fit for the recreational purposes of colonial societies. Understood in this manner, turf grass monoculture is cast in the role of a nonhuman vector of the settler-invader state: an entanglement between the human and nonhuman aspects of the imperial project that is laid over what Carter refers to as the "lost surfaces of the [pre-colonial] world."[19] Carter's ecocritical model thus provides a template for the assertion that the lawn is not just akin to colonialism, but an aspect of it: an attempt to assert an ecological arrangement upon a locale. The lawn is a form of ongoing "natural" oppression: the nonhuman manifestation of a casually cruel and mindlessly monolithic cultural template imposed upon the world that provides a dire prediction for the wider role of human agency in the Anthropocene. However, whereas for Carter this transformation ushers in an "embrace of environmental amnesia," the actual business of lawn maintenance suggests that forgetting the world that came before is not a privilege afforded to most cultivators of the lawn.

THE NONHUMAN WRITES BACK

As anyone who has been responsible for a patch of lawn will realize, they do not permit much in the way of forgetting: either literally or figuratively. Those who forget to tend to their lawn are reminded fairly quickly of the potential for the nonhuman to escape human control and thrive and expand in unforeseen ways. Unwatched grass can grow surprisingly quickly. Such material burgeoning speaks also to larger questions regarding the impossibility of ever "forgetting" the ecological arrangements that existed before or permanently erasing all vestiges of prior ecologies. Previous environments are not excised through the establishment of a lawn, and have ways of reasserting themselves. Depending on the location, plant species can establish themselves without invitation; unanticipated fauna find a home above or below the soil; water becomes either scarce or overabundant; temperatures soar or plunge in ways that threaten the viability of turf grass; the ground itself can subside or expand under the influence of tree roots or tectonic plates. Testament to the difficulties of lawn maintenance, the historical success of the lawn in different ecological contexts has been far from assured. Australia, the "dry continent," does not always afford the irrigation needed for grass growth[20] and the continental United States spans a wide range of climactic zones, many of which are not conducive to long-term grass cultivation.[21] No lawn is simply a product of human assertion, but rather arises in its specificity through the co-authorship of the local nonhuman: a perspective that acknowledges "a natural environment whose material forces themselves manifest certain agentic capacities and in which the domain on unintended or unanticipated effects is considerably broadened."[22] The nonhuman is not dead matter that simply acquiesces to the idea of the lawn and thus such environmental arrangements require sustained human intervention and engagement to be maintained. While the brute facts of lawn inputs are often read as a sign of comfortable human dominance, this is far from the only possible interpretation. The intensity of such interventions can also be read as a case of human desperation in the attempt to exert influence upon spaces that resist and reject such human-enforced ecological formations.

The unstoppable agency of the local lawn points towards the limitations of lawn critique as expressed by figures such as Pollan, especially insofar as his opposition to lawn mowing recalls the mass culture tradition discussed earlier. Lawn critique and the mass culture critique share a common methodological blind spot in their pre-emptive dismissal of even the possibility of heterogeneity. Such an omission is particularly telling when advocated for by those with an exceptionally rich conceptual framework for analysing and indeed celebrating specificity. For example, assessing the limitations of the methods of the sometimes mass culture critic, Theodor Adorno, Fredric Jameson argues that his:

> [T]endency to dismiss all of manipulated mass culture as false consciousness … leaves unused and deprives of their very reason for being the very rich instruments of formal analysis also developed by Adorno, which in his work only come to find their fullest application in the decipherment and deconcealment of the formal contradictions at work in the monuments of high culture[23]

In effect, Jameson is arguing that Adorno's pre-emptive dismissal of mass culture as unworthy of close analysis means that his conclusion is never tested: because he begins with the assumption that mass culture is aesthetically bankrupt, he can dismiss all examples without consulting any of them. The logic of this argument equally applies in the context of lawns debates: where the actual existence of individual lawns is passed over in favour of a denouncement of the abstract idea of the ideal lawn and its implicit conformity. As Adorno deemed mass culture unworthy of his carefully cultivated attention, so the lawn is deemed unworthy by those critics who have carefully honed their ability to analyse and understand the ecological details of other environments. Like mass culture critics, lawn critics very rarely dirty their hands with actual examples. For example, early in his essay, Pollan notes how lawn mowing brought him into a deep connection with the lawn space, where he "knew by heart the exact location of every stump and stone, the tunnel route of every resident mole, the address of every anthill." However, by the time he moves into his rejection of the lawn, the specificity is gone: it is nothing more than brute monotony and the lawn is reduced to nothing more than a site of brute imposition. In a manner that mirrors the dismissal of popular culture, it is not that the lawn lacks complexity and possibility, but that Pollan has decided to shut himself of from it. It is this specificity, or lack thereof, that drives Pollan's critique and indeed informs the more apocalyptic manifestations of lawn critique more generally.

Pollan's nightmare dream of a monolithic lawn that stretches on forever, never changing, is just that: a dream. At a remove, the lawn appears a perfect manifestation of the lush and even "lawn aesthetic": a fantasy promoted by chemical-input industry to increase demand for commercial lawn care products.[24] However, the endless, changeless lawn cannot exist in practice regardless of the intensity of human inputs and action in the lawn space due to the constant disruption and agency of the nonhuman. In contrast to this abstract vision of endless lawn monotony, a consideration of specific manifestations of lawn aesthetics illuminates the particular fears around difference and totality that shape our engagement with the nonhuman. In practice, all lawns are different: shaped by very precise climactic, botanic, tectonic and zoological force. Lawn critique's vision of dull repetition is premised on an assumption of the surrounding environment as dead matter, rather than an assemblage of active, vital and productive co-authors that compels us "to consider anew the location and nature of capacities for agency."[25] It is when we forget that the lawn is always local in its manifestation, its meanings and its entanglement with the nonhuman, that it appears irredeemable and alien. Consideration of the multiple scales thus opens up the possibility that the uniform lawn aesthetic can be interrupted by instances of disjuncture, in the sense originally developed by Arjun Appadurai,[26] where the homogenous application of the idea (and the ideal) of the lawn is disrupted through its interaction with local environmental conditions. In Appadurai's terminology of "landscapes," global forces are not imposed homogenously across the world but rather manifestly different in different places as they come into contact with other formations, flows and agents which he categorizes in terms of five "-scapes." Taking up his terminology, we can cast the global spread of lawn culture as constituted by encounters between the "ideoscape" of the lawn – "a chain of ideas, terms, and images" – coming up against local environmental conditions, which we might name through a new term: the "ecologoscape." Although different colonial nations (as well as other forms of state) share a common vision of a lawn, the ability to realize this vision is always subject to the influence of the nonhuman peculiarities of particular places. The antidote to homogenous lawn critique is to be always attentive to actual lawns from which the larger system arises, while not losing sight of the multiple scales on which lawn environments operate: individual, community and environment.

Such an approach allows us to move beyond the assumption that every lawn aspires to hegemonic uniformity, and instead consider them as spaces "for socially and ecologically rich and engaging alternatives."[27] One example of such thinking is Bormann et al.'s notion of the "freedom lawn" a lawn that is mowed, but not weeded and thereby "results from an interaction of naturally occurring processes and the selective efforts of lawn mowing."[28] While somewhat banal, especially for those unaccustomed to complimenting lawn mowing with chemical regimes, the freedom lawn marks a practical example of a lawn where nonhuman authorial agency is actively acknowledged. Similarly, Head and Muir address lawns and backyards as spaces where the human and nonhuman coexist because "nature has a life of its own and can act back, or in any other direction": It is through engaging with such backyard spaces that human actors can come to know and appreciate the "nature" of the local nonhuman.[29] Such specific examples share a common and practical attentiveness to both the idea of the lawn as a communal project and the power of the nonhuman to shape the lawn space. They thus both alert us to the absence of the nonhuman in apocalyptic lawn critique, which therefore tends to overstate the destructive power of lawnmower, and demonstrate how the nonhuman works as a physically and conceptual counter to the conformist paradigm of lawn thinking. When the nonhuman is taken into account as an inevitably active participant, and not simply a passive victim, then we cannot help but think about the lawn locally, because that is one of the key scales at which the nonhuman operates. The aesthetic existence of actual lawns are thus a product of the inevitable existence of difference and variety that emerges out of the intersection of wider cultural forces and particular nonhuman forces in the context of lawn spaces. The celebration of such lawn forms demonstrates how concern for aesthetic relations with the nonhuman need not inevitably load to domination and destruction, but can also involve mutually meaningful entanglements.

The lawnmower thus functions as a technological hinge that facilitates engagement and conversation between the human and nonhuman.[30] From the perspective of lawn critique, this hinge exerts a double act of conformity: enforcing both nonhuman conformity, by reducing every blade of grass to a single height, and enabling a technological and social system that compels human agents to engage in identical acts at the behest of the larger group. Conceived as the exemplary instrument of the suburbs, the lawnmower exerts a homogenizing influence along both human and nonhuman axes simultaneously and thereby reinforces their separation from one another. However, when we shift from the suburban-cultural fear of the totalizing lawn aesthetic to an approach that is both more materially aware and aesthetically nuanced, the lawnmower emerges as a point of articulation that can act to bridge the gap. Rather than a tool of total uniformity, the lawnmower can instead be considered an instrument by which human agents intervene and shape the local environment in ways that bring awareness of the specificities of that space. As a means of intervention into the nonhuman world, the lawnmower facilities interaction and conversation, not simply imposition. Through the act of mowing, human actors come to know the local landscape in time (what species appear at certain times of the year? how does climate alter plant growth? how does the earth subside and rise?) and space (what are the contours of the land? how does the nonhuman move across human boundaries?). Listening is important in this engagement, but it is not the full measure of it, because this engagement also involves human action and agency as well: lawn mowing teaches us that we should not be afraid to say and do things to the nonhuman provided we are willing to listen and observe what happens in response. The lawn reminds us that it is not only dreams of "wild nature" that can inform our ecological imaginations: the nonhuman continues to speak to us even after – especially after – we have engaged with it in a material sense.

CONCLUSION

The lawn is an aesthetic space, but one that is always co-authored by the nonhuman, as well as the human. In their aesthetic existence, actually existing lawns need to be understood as the product of multiple forces operating on multiple levels: the globally circulating idea of the lawn, the felt obligation to local community, the active intervention and entanglement of the nonhuman and the agency of the individual human actor pushing a rapidly rotating blade across a patch of grass. In contrast to its apparent mundanity, the lawn thus emerges as the site of multiple contentious and complicated but also largely intentional entanglements between human and nonhuman actors, community responsibility and private expression, local conditions and global ideas. The mutually produced project of the lawn is never complete, but rather a never-ending site of entanglement and conversation where human actors neither recoil from nor apologize for their involvement in nonhuman spaces, but instead own their involvement in ways that can provide a basis for reflection and judgement. This is where lawn mowing can provide a productive model for thinking beyond the current situation, where most human interventions in the nonhuman world are largely unintentional, unforeseen and accidental. Given the constellation of forces at play, the debates around the practice of lawn mowing replicate on a smaller-scale current conflicts about how to manage human influence in the Anthropocene: in particular the inescapable fact of human agency in the nonhuman world. If, in the Anthropocene, human actions are going to have environmental implications, then lawn mowing is one possible guide for how we might consciously address and debate the nature of those actions and implications and thereby responsibly navigate the ethical and political repercussions of a world where humans and nonhumans are inextricably entangled. What therefore might be gained from scaling up the lessons of the lawn to the global context of the Anthropocene?

Perhaps the most important lesson we can draw from lawn mowing in the current discussion is that where human action is involved we cannot restrict our debates to the rational determination of practical use. Humans are more than practical animals: we are equally expressive and aesthetic beings. Questions of aesthetics are therefore more than simply distractions in human engagements with the nonhuman; they are a necessary and inescapable aspect of the Anthropocene. Consequently, it is insufficient to declare that "human aesthetics rarely correspond to ecological principles" and then seek to privilege the later over the former,[31] because such a conclusion denies the intractable importance of aesthetics to human existence and flourishing. We cannot completely deny aesthetics because they conflict with ecologically correct living, because to do so would to produce a world antithetical to meaningful ways of life. Rather than deny aesthetics, our goal should be to consider how aesthetic priorities might be accounted for, negotiated with and even aligned with the ecological priorities of the Anthropocene. Lawn mowing provides us with the resources for doing this because it constitutes a widespread form of ecological entanglement where many of us already actively engage with and shape the nonhuman world: for all but the most anti-social or lazy of lawn curators, the involvement of human action and agency is taken as inescapable element of the lawn space. Decisions about lawn mowing are therefore never simply isolated debates to be decided on purely ecological grounds, but always intersect with economic, aesthetic and other ideological forces that can be equally legitimate and meaningful.

In particular, lawn mowing offers a model for thinking through the role of aesthetic engagements with nonhuman forces in the context of the Anthropocene. Lawn critique manifests an aesthetic form of deep ecology: one where nature, following Bruno Latour, is not only out of reach of political deliberation,[32] but also of aesthetic judgement, and the most beautiful and desirable ways of arranging the world are those where humanity plays no part. However, contrary to such a model, the actual practice of lawn mowing embodies an aesthetically sensitive engagement with the nonhuman that is not determined by conservationist assumptions. Lawn mowing thus has the effect of reopening up that closed space: emboldened by its small scale, the lawnmower dares to intervene and reapportion the arrangement of the nonhuman and thereby reveals that such engagement does not utterly destroy it, but instead opens up new ways of living and working with it. Moreover, from such small-scale, apparently individual efforts, larger change can be effected: although maintained on the local level, lawns have continental reach: in the United States, it is estimated that lawns are now the "largest irrigated crop" by land area, covering 128,000 square kilometres.[33] Lawn mowing is not a war against nature, or at least it need not be, it can also be a collaboration, a duet with the nonhuman world where we build new natural or post-natural spaces that express and meet human needs, but are no less natural for doing so. Lawns are a space in which many engage with the nonhuman world in a precise and attentive manner that is otherwise unavailable to them and in doing so both enjoy and better understand their implication in the nonhuman world. In this manner, lawn mowing teaches us that, in aesthetic terms at least, conscious and directed engagement with the nonhuman need not be conceived of as destruction of nature, but rather the transformation of one arrangement of the human and nonhuman worlds into another.[34] On the larger scale of the Anthropocene, lawn mowing thereby provides a model for making sense of the different ways in which human agency – understood on both collective and individual levels – can produce more, as well as less, desirable forms of entanglement with the nonhuman.

In the Anthropocene, humanity's involvement and influence over the nonhuman is inescapable. Coming to terms with the possible consequences of human agency is therefore of central importance, because the idea of nature can no longer be "wheeled out to adjudicate between what is fleeting and what is substantial or permanent."[35] Most frequently, that influence is understood as almost inevitably damaging for both humans and nonhumans. However, lawn mowing provides a context for evaluating other outcomes of these encounters. Lawn mowing debates thus provide fertile terrain for understanding how humans can responsibly and productively engage with the nonhuman world. Rather than advocating for withdrawal or minimization of impact upon the nonhuman as the only responsible course of action, a model of entanglement based on lawn mowing suggests the possibility – but not the inevitability – of positive and enriching forms of human intervention and influence. By opening up the potential for desirable outcomes, lawn mowing calls on us not to renounce or relinquish humanity's ability to influence the nonhuman (as if that were even possible) but instead to take responsibility for humanity's agency, ability and compulsion to transform the nonhuman. Certainly, the lawn is not a perfect role model: lawns can be extremely destructive, especially when maintained in environments, such as the American Southwest, where turf grass has extreme difficulty simply surviving. However, even in such moments of failure, the lawn is nonetheless instructive, because such failures call on us to recognize our agency and respond accordingly. Thus, increasingly in places like California we are witnessing a transformation in lawn thinking, with the introduction of species that are better suited to thrive in local environments.[36] Moreover, in other places, recent ecological research suggests that, in stark contrast to the dark dreams of lawn critique and mass nature, environments dominated by turf grass can support a wide variety of nonhuman life[37] (although species that thrive in such environments are often vilified and denounced as less worthy manifestations of the nonhuman[38]). In such contexts, intentional engagement with the nonhuman via the lawn has facilitated new conversations and new knowledge about human entanglement with the environment at a local level with the potential for larger consequences. In this manner, lawn mowing provides a case study of how human and nonhuman entanglements can be implemented, assessed, rejected and sometimes embraced: a model for how we might learn to think and to live in the context of the Anthropocene.

NOTES

1. Lesley Head and Pat Muir, *Backyard* (Wollongong: University of Wollongong Press, 2007), 64.
2. Paul Robbins and Julie Sharp, "Turfgrass Subjects: The Political Economy of Urban Monoculture," in *In the Nature of Cities*, ed. Nik Heynen, Maria Kaika and Erik Swyngedouw (London: Routledge, 2006), 120.
3. Paul Robbins and Julie Sharp, "Producing and Consuming Chemicals: The Moral Economy of the American Lawn," *Economic Geography* 79:4 (2003), 428–32; Paul Robbins and Julie Sharp, "The Lawn-Chemical Economy and its Discontents," *Antipode* 35:5 (2003): 957–70.
4. Jennifer Blecha, "Lawns," in *Sage Encyclopaedia of Environment and Society*, ed. Paul Robbins (Thousand Oaks: Sage, 2007), 1031.
5. Virginia Scott Jenkins, *The Lawn: A History of an American Obsession* (Washington: Smithsonian Institution Press, 1994), 28–30.
6. F. Herbert Bormann, Diana Balmori and Gordon T. Geballe, *Redesigning the American Lawn: A Search for Environmental Harmony* (New Haven: Yale University Press, 1993), 23.
7. Robbins and Sharp, "Producing and Consuming Chemicals," 442.
8. Elizabeth M. Cook, Sharon J. Hall and Kelli L. Larson, "Residential Landscapes as Social-Ecological Systems: A Synthesis of Multi-Scalar Interactions between People and their Home Environment," *Urban Ecosystems* 15 (2012): 37.
9. Robbins and Sharp, "Producing and Consuming Chemicals," 445.
10. Albert Way, "Review of *American Green* and *Lawn People*," *Geographical Review* 98:4 (2008): 576.
11. Robert Messia, "Lawns as Artifacts: The Evolution of Social and Environmental Implications of Suburban Residential Land Use," in *Suburban Sprawl: Culture, Theory and Politics*, ed. Matthew J. Lindstrom and Hugh Bartling (Lanham: Rowman & Littlefield, 2003), 69–70.
12. Franklin Ginn, "Sticky Lives: Slugs, Detachment and More-than-human Ethics in the Garden," *Transactions of the Institute of British Geographers* 39 (2013): 533.
13. Dean MacCannell, "Landscaping the Unconsciousness," in *The Meaning of Gardens*, ed. Mark Francis and Randolph T. Hester Jr. (Cambridge: MIT Press, 1991), 94.
14. Head and Muir, *Backyard*, 65.
15. Jenkins, *The Lawn*, 133–35.
16. Dwight MacDonald, "A Theory of Mass Culture," in *Mass Culture: The Popular Arts in America*, ed. Bernard Rosenberg (Glencoe, IL: Free Press, 1957), 62.
17. Blecha, "Lawns," 1030; Bormann, Balmori and Geballe, *Redesigning the American Lawn*, 13–28; Paul Robbins, *Lawn People* (Philadelphia: Temple University Press, 2007), 30.
18. Paul Carter, *The Lie of the Land* (London: Faber and Faber, 1996), 2.

19. Ibid., 4.

20. Head and Muir, *Backyard*, 124.

21. Bormann, Balmori and Geballe, *Redesigning the American Lawn*, 30–31.

22. Diana Coole and Samantha Frost, "Introducing the New Materialisms," in *New Materalisms: Ontology, Agency, and Politics*, ed. Diana Coole and Samantha Frost (Durham: Duke University Press, 2010), 10.

23. Fredric Jameson, *Valences of the Dialectic* (London: Verso, 2009), 334.

24. Robbins and Sharp, "Producing and Consuming Chemicals," 434–36.

25. Coole and Frost, "Introducing the New Materialism," 9.

26. Arjun Appadurai, *Modernity at Large: Cultural Dimensions of Globalization* (Minneapolis: University of Minnesota Press, 1996), 37.

27. Robbins and Sharp, "The Lawn-Chemical Economy and its Discontents," 974.

28. Bormann, Balmori and Geballe, *Redesigning the American Lawn*, 57.

29. Head and Muir, *Backyard*, 8.

30. I would like to thank the anonymous reviewer for offering this incisive and powerful insight.

31. Kerry J. Dawson, "Nature in the Urban Garden," in *The Meaning of Gardens*, ed. Mark Francis and Randolph T. Hester, Jr. (Cambridge: MIT Press, 1991), 143.

32. Bruno Latour, *The Politics of Nature*, trans. Catherine Porter (Cambridge: Harvard University Press, 2009), 26–27.

33. Blecha, "Lawns," 1031.

34. Paul Robbins and Trevor Birkenholtz, "Turfgrass Revolution: Measuring the Expansion of the American Lawn," *Land Use Policy* 20 (2003): 181–94.

35. Timothy Morton, *Ecology without Nature: Rethinking Environmental Aesthetics* (Cambridge: Harvard University Press, 2007), 21.

36. Jennifer Medina, "In Drought-Ridden California, the Classic Lawn Loses Ground," *New York Times*, 26 October 2015. Online.

37. Johnny Cornelis and Martin Hermy, "Biodiversity Relationships in Urban and Suburban Parks in Flanders," *Landscape and Urban Planning* 69 (2004): 398–99; Stephen DeStefano and Richard M. DeGraff, "Exploring the Ecology of Suburban Wildlife," *Frontiers in Ecology and the Environment* 1:2 (2003): 95–101; Mark A. Goddard, Andrew J. Dougill and Time G. Benton, "Scaling Up from Gardens: Biodiversity Conservation in Urban Environments," *Trends in Ecology and Evolution* 25:2 (2009): 90–98; Michael L. McKinney, "Urbanization, Biodiversity, and Conservation," *BioScience* 52:10 (2002): 885–89; Jean-Pierre L. Savard and J. Bruce Falls, "Survey Techniques and Habitat Relationships of Breeding Birds in Residential Areas of Toronto Canada," in *Avian Ecology and Conserviation in an Urbanizing World*, eds. John Marzluff, Reed Bowman and Roarke Donnelly (Boston: Kluwer Academic Publishers, 2001), 559.

38. Nicholas Holm, "Consider the Squirrel: Freaks, Vermin and Value in the Ruin(s) of Nature," *Cultural Critique* 80 (2012): 56–95.

BIBLIOGRAPHY

Appadurai, Arjun. *Modernity at Large: Cultural Dimensions of Globalization*. Minneapolis: University of Minnesota Press, 1996.

Blecha, Jennifer. "Lawns." *Sage Encyclopaedia of Environment and Society*. Ed. Paul Robbins. Thousand Oaks: Sage, 2007, 1029–31.

Bormann, F. Herbert, Diana Balmori and Gordon T. Geballe, *Redesigning the American Lawn: A Search for Environmental Harmony*. New Haven: Yale University Press, 1993.

Carter, Paul. *The Lie of the Land*. London: Faber and Faber, 1996.

Cook, Elizabeth M., Sharon J. Hall and Kelli L. Larson. "Residential Landscapes as Social-Ecological Systems: A Synthesis of Multi-Scalar Interactions between People and their Home Environment." *Urban Ecosystems* 15 (2012): 19–52.

Coole, Diana and Samantha Frost. "Introducing the New Materialisms." *New Materialisms: Ontology, Agency, and Politics*. Ed. Diana Coole and Samantha Frost. Durham: Duke University Press, 2010.

Cornelis, Johnny and Martin Hermy. "Biodiversity Relationships in Urban and Suburban Parks in Flanders." *Landscape and Urban Planning* 69 (2004): 385–401.

Dawson, Kerry J. "Nature in the Urban Garden." *The Meaning of Gardens*. Ed. Mark Francis and Randolph T. Hester, Jr. Cambridge: MIT Press, 1991, 138–43.

DeStefano, Stephen and Richard M. DeGraff. "Exploring the Ecology of Suburban Wildlife." *Frontiers in Ecology and the Environment* 1:2 (2003): 95–101.

Ginn, Franklin. "Sticky Lives: Slugs, Detachment and More-than-human Ethics in the Garden." *Transactions of the Institute of British Geographers* 39 (2013): 532–44.

Goddard, Mark A., Andrew J. Dougill and Time G. Benton. "Scaling Up from Gardens: Biodiversity Conservation in Urban Environments." *Trends in Ecology and Evolution* 25:2 (2009): 90–98.

Head, Lesley and Pat Muir. *Backyard*. Wollongong: University of Wollongong Press, 2007.

Holm, Nicholas. "Consider the Squirrel: Freaks, Vermin and Value in the Ruin(s) of Nature." *Cultural Critique* 80 (2012): 56–95.

Jameson, Fredric. *Valences of the Dialectic*. London: Verso, 2009.

Jenkins, Virginia Scott. *The Lawn: A History of an American Obsession*. Washington: Smithsonian Institution Press, 1994.

Latour, Bruno. *The Politics of Nature*. Trans. Catherine Porter. Cambridge: Harvard University Press, 2009.

MacCannell, Dean. "Landscaping the Unconsciousness." *The Meaning of Gardens*. Ed. Mark Francis and Randolph T. Hester, Jr. Cambridge: MIT Press, 1991, 94–102.

MacDonald, Dwight. "A Theory of Mass Culture." *Mass Culture: The Popular Arts in America*. Ed. Bernard Rosenberg. Glencoe, IL: Free Press, 1957, 98–107.

McKinney, Michael L. "Urbanization, Biodiversity, and Conservation." *BioScience* 52:10 (2002): 883–90.

Medina, Jennifer. "In Drought-Ridden California, the Classic Lawn Loses Ground." *New York Times*, 26 October 2015. Online.

Messia, Robert. "Lawns as Artifacts: The Evolution of Social and Environmental Implications of Suburban Residential Land Use." *Suburban Sprawl: Culture, Theory and Politics*. Ed. Matthew J. Lindstrom and Hugh Bartling. Lanham: Rowman & Littlefield, 2003, 69–84.

Morton, Timothy. *Ecology without Nature: Rethinking Environmental Aesthetics*. Cambridge: Harvard University Press, 2007.

Robbins, Paul. *Lawn People*. Philadelphia: Temple University Press, 2007.

Robbins, Paul and Trevor Birkenholtz. "Turfgrass Revolution: Measuring the Expansion of the American Lawn." *Land Use Policy* 20 (2003): 181–94.

Robbins, Paul and Julie Sharp. "The Lawn-Chemical Economy and its Discontents." *Antipode* 35:5 (2003): 955–79.

———. "Producing and Consuming Chemicals: The Moral Economy of the American Lawn." *Economic Geography* 79:4 (2003): 425–51.

———. "Turfgrass Subjects: The Political Economy of Urban Monoculture." *In the Nature of Cities*. Ed. Nik Heynen, Maria Kaika and Erik Swyngedouw. London: Routledge, 2006, 110–42.

Savard, Jean-Pierre L. and J. Bruce Falls. "Survey Techniques and Habitat Relationships of Breeding Birds in Residential Areas of Toronto Canada." *Avian Ecology and Conservation in an Urbanizing World*. Ed. John Marzluff, Reed Bowman and Roarke Donnelly. Boston: Kluwer Academic Publishers, 2001, 543–68.

Way, Albert. "Review of *American Green* and *Lawn People*." *Geographical Review* 98:4 (2008): 576–79.

Chapter 3

Learning with the River

On Intercultural Gifts from the Whanganui

Charles Dawson

E rere kau mai te Awa nui mai te Kahui Maunga ki Tangaroa
Ko au te Awa, ko te Awa ko au
The great River flows from the mountains to the sea
I am the River and the River is me

—Māori adage from the Whanganui River

Fresh water shapes and nourishes ecological and cultural entanglements. The indigenous Māori relationship with water makes it imperative that New Zealand formulates policy responses alive to "the cultural meanings encoded in water."[1] This is occurring: the country's longest navigable river (or *awa*), *Te Awa o Whanganui*, will have standing: it will gain formal recognition before the institutions that have previously dammed, and diverted its waters. The legislation acknowledging the River's customary and spiritual *mana* (prestige and authority) is the fruit of 125 years of petitions, litigation, claims and protest from Māori, derived from the settlement of claims under the Treaty of Waitangi 1840.[2] This Treaty settlement is one of the few examples of a river's physical and metaphysical elements being recognized in modern law. The concept is a profound gift from Māori. Settlements negotiated between the Crown (government) and Māori demand a wider *Pākehā* (white) understanding of Māori world views. The agreement's inclusive approach fosters a bicultural, relational approach to justice, environment and reconciliation.

This chapter is informed by my father's work alongside Māori for recognition of their Treaty rights. It retells one Pākehā (white) story, drawing on public statements about the relationship between Māori and their *Awa*, and what the River's new status might ask of *Pākehā* New Zealanders. In the law, ideas of ownership are replaced by notions of *kaitiakitanga*, or guardianship, and traditional Māori concepts shape regulatory structure with real elegance. The challenge is to extend the conversation beyond those who have fought for or been captivated by the River, to encompass gratitude, humility and deeper change within New Zealand's dominant neoliberal paradigms predicated on exploitation of resources. The Whanganui emerges as both guide and vessel, burdened and enabled by its histories.

The life and career of my father, Martin Dawson, was transformed by his work in partnership with Māori for recognition of their rights under the Treaty of Waitangi. The work was extensive, instructed by leaders in areas such as language revitalization, broadcasting, fisheries, forestry, water rights, cultural and intellectual property, and the foreshore and seabed, in a series of important cases that slowly redefined New Zealand's legal and policy approaches to the Treaty. For him the law was a tool for justice. The corpus of New Zealand scholarship at the intersection of indigenous rights, customary practice, law, history (written and oral), reparation and reconciliation has expanded rapidly since the late 1980s, but when Martin began the work in 1983, the disciplinary entanglement inherent in Treaty negotiations and hoped-for reconciliation was in its nascent form.

The Treaty of Waitangi, signed by representatives of the British Crown and Māori in 1840, promised Māori *tino rangatiratanga* or full and exclusive control over their lands and culture and a decisive role in the formation of the new nation; for Māori the Treaty was to be a unifying compact. The Treaty's provisions were rarely followed, particularly after the major battles of the New Zealand Wars of the 1860s–1870s. The unprecedented speed of colonization and forest clearance in the following five decades ruptured Māori biocultural worlds. The Waitangi Tribunal saw an opportunity lost:

> The settler mindset appeared to be that communal values were fundamentally incompatible with progress. The narrow view taken by those in authority meant that they saw only one solution – Māori law and administration had to be totally replaced with settler laws.
>
> As a consequence, an evolution that occurred over centuries in Britain was squeezed into just a few decades in New Zealand. … These changes lead to Māori and Pākehā talking past each other to a much greater degree than is willingly acknowledged.[3]

The rapid pace of change led to a form of mis-remembering. The Treaty was ignored, largely forgotten in official storage, to suffer from rodent and water damage. Recourse to the Treaty was regarded with bafflement, disdain or "benign neglect" by many in the law or government until the 1980s.[4] The country thought it had the best race relations in the world, until Māori grievances – and their generosity with their stories – countered the complacency. From the 1970s, activist Māori looked to their elders but also to the Black Panther and AIM movements, protesting for Treaty rights and government action, along with some appearances before the quasi-judicial truth and reconciliation commission: the Waitangi Tribunal. In 1985, the Tribunal was empowered to inquire into Treaty breaches back to 1840.[5] Inquiries unearthed broken or neglected Treaty promises, wayward good intentions, and the "legalised rapacity" of government land acquisition.[6]

My father's clients led him to see the potential of the Treaty as both healing force and guiding instrument. He worked on the watershed 1987 *Lands* case which found, "The duty of the Crown is not merely passive but extends to active protection of Maori people in the use of their lands and waters to the fullest extent practicable."[7] The government had to re-evaluate its relationship with Māori, and to understand (and then, constrain), its Treaty obligations. In the months around my mother's death, Martin was swamped with work.

* * *

My father sits at his office desk, smoothing his brow with his left hand, as he edits a memorandum. Clearing his throat, he stands, picks up his dictaphone, rewinds the tape, and, consulting his notes amidst a mountain of files, he begins to talk. He falls into the dictaphone user's absorbed pace, Māori terms weaving their way into his clear, forceful legal language. The legal phrasing is peppered by phrases from the Treaty of Waitangi, such as tino rangatiratanga, *the full exclusive and undisturbed control promised in Article Two of the Treaty, or* taonga, *those treasured cultural elements (such as language, and those places, practices, flora and fauna, and associated knowledge which Māori understood the Crown would protect through a collaborative Treaty partnership). I'd often seen that dictaphone intensity, watching him pace the floor and navigate the growing piles of papers, books and folders on his desk. Ever since he got swept up in the intensive Treaty work, any stray page would become home to notes about a case. The work had ecological and cultural entanglement at its core, and it caused alarm and consternation. Pacing by those papers, Martin kept his hand on the tiller.*

* * *

Martin was the first to acknowledge just how much he learned about New Zealand's bicultural potential from his many guides, supporters, colleagues and mentors. When he was invited to represent Māori of the Whanganui River, I also met those who speak and listen to *Te Awa* as a core part of who they are; those people transmit lessons from that inherent ecological (and specific communal) relationship with those storied waters. Many Waitangi Tribunal reports and other sources explicate Māori water concepts. Te Ahukaramū Charles Royal explains:

> *Te taha wairua* can literally be translated as "the dimension of two waters," a conception that likens spirituality to water. However, it might be argued that *te taha wairua* does not mean "the spiritual plane" at all. Instead, references to *te taha wairua* might be saying that there is a fundamental dimension to all life and it takes the form of water.[8]

Linda Te Aho notes the Māori understanding of *wai* (water) does not end at the water's edge:

> Rivers are often conceptualised as living beings, ancestors with their own *mauri* (life force), *mana* (prestige and authority) and *tapu* (sacredness). As such they are whole and indivisible entities, not dissected into beds, banks and water not into the tidal and non-tidal, navigable and non-navigable parts. The saying "*ko au te awa, ko te awa ko au*" (I am the river and the river is me) comes from the Whanganui River, but speaks to this interconnectedness that lies at the heart of the way many Māori view the world and our waterways.[9]

The Māori understanding of water scrutinizes the healthy flow of essential life energy, or *mauri*. Healthy ecosystems allow species and *mauri* to flourish; stories and traditions of relationship with the species flow too, within the practice of *kaitaiakitanga* or guardianship fostered in a revival in tribal journeying for modern youth.

TE TIRA HOE WAKA: LISTENING TO THE WHANGANUI

The Whanganui River's headwaters gather on the slopes of the sacred mountains Tongariro and Ruapehu. The *Awa* then flows 180 miles southwest from the central North Island, reaching the sea at the city of Whanganui. There are some 240 rapids along its course, and each one can teach the paddler. Tribal connections with the River were and are paramount. Every January since 1989, Māori of the River have gathered to make a renowned canoe journey – the *Tira Hoe Waka* or company of canoes – along their ancestral current. The *Tira* (travelling party) affirms the traditions and connections the River nurtures. Two weeks immersed in prayer, ritual, laughter, song, paddling and teamwork, from the mountains to the sea, exposes *rangatahi* (youth) to their history: "'at each rapid, *kaitiaki* and people dwell.' We have been taught that the *kaitiaki* [guardian beings] within the waterways, upon the mountains – wherever in the tribal domain they reside – are our ancestors. Therefore, we do not speak about the *Awa*, instead we commune directly with the *Awa*."[10]

* * *

It's 1990, and I am paddling on the Whanganui River. Archie Taiaroa, a River leader, invited me on this journey after I volunteered on aspects of the tribe's claims. Across two weeks immersed in prayer, traditional knowledge, discussion and laughter (as when people tip out in famous rapids like Ngāporo*), the journey honors the past and present. Learning by doing, we understand there are several ways of knowing water. It was through* te Tira Hoe Waka, *my father, and my stepmother (the anthropologist Patricia Laing) that I first met those learned Māori steeped in traditional River lore. These elders enriched the lives of all those who knew them. Their dignity, their palpable and electrifying* mana *(dignity, presence, stature), their* aroha *or love: these qualities flowed from them, changing and challenging those they met. These elders knew what* Te Awa *could offer and were masterful in imparting the abundant spirit of the place, and the place of the spiritual in the world. The journey transmits histories and traditions of* te Awa *to Māori descendants, binding specific cultural identities to certain rapids or places, consolidating whakapapa, the human and nonhuman genealogy affirmed in many ways.*

The people of the River and the River itself are the real teachers. The Whanganui will call up what you have submerged, but it will also guide you. The Māori saying I ngā rā ō mua *(the past is before you), applies as the river current unfurls beneath the hull of a canoe. On the* Tira Hoe Waka, *I paddled with others in a once-buried canoe, torqued up from the depths after decades beneath the eddies. My understanding of water and cultural identity was transformed. Gratitude and indebtedness are bound here: Māori of the river shared their insights and showed me their river, waiting for the powers that be to stop and listen. Ritual, song, and the recognition of traditional protocols are crucial in the journey. So were spans of silence and reflection, droplets falling off the paddle into the flow, the stars riffling in the upriver eddies, and the presence of guardian spirits observing us as we passed.*

* * *

Māori recognition of the currents of memory is not about some sort of riparian daydream, nor is the connnection something *Pākehā* can neglect or disdain any longer. Whanganui leader and former politician Dame Tariana Turia says the journey "is a vehicle for our survival" against (neo)colonial systems designed with Pākehā world views in mind.[11] The gathering has nurtured a new generation of leaders at the highest levels, who are bearing their elders' vision and passion into law and policy: "The *Tira Hoe* is marked by memory, a concentration upon and reconstitution of the past, and the desire of the elders to transmit to a younger generation the importance of their heritage."[12] Listening to the river – "thinking like a river" as Donald Worster put it – matters.[13] Māori share their understanding of what Ivan Illich (*pace* Gaston Bachelard) termed "the fluid that drenches the inner and outer spaces of the imagination."[14] Paddling into the past, being controlled by that River-as-ancestor, can become a healing act.[15] There are generations of hurt to paddle through. Acknowledging Treaty breaches means the government's practice of apologizing to the tribes central to healing and renewing relationships: it facilitates the beginnings of the healing border work of empathy and remembering the river: "in bridging past and present, collective apologies counteract those who find it politically convenient to keep them separate."[16]

A BEHEADED EEL: THE WHANGANUI AND THE LEGACY OF HYDRO SCHEMES

For Ngāti Rangi, the tribe at the headwaters, and Te Ātihaunui-a-Pāpārangi, the peoples downriver, *Te Awa* is life way and anchor, its storied pulse reflecting tribal and river health. Keith Wood, of Ngāti Rangi, speaks of the connections between ancestors, place, waters and the transmission of guardianship practices:

> We believe we "belong" to these ancestral waters. We consider ourselves to be an integral part of the environment as *whanaunga* (relations). This connection is also the basis for our relationship of *kaitiakitanga* [guardianship and interconnection]. It is from this relationship that our role as *tangata tiaki*, as guardians and caregivers of our *awa*, comes from. … The meaning of our world lies in our relationships. … Our values emphasise the interconnectedness of the natural world and people.[17]

Belonging to the waters. Keith Wood's world view contends with the concrete interventions of the dams that dominates New Zealand rivers: "irrigation and electricity demands on our free-flowing waters have always had their way."[18] Hydro-electricity schemes offer an impressive 54 percent of New Zealand's power network.

In the 1950s, the modernist might of hydro-engineering came up against the Māori spiritual and cultural legacy of the Whanganui River, and ignored it. From 1958, plans were made for an ambitious hydro-engineering scheme, overseen by London engineers and public works staff. In 1974, the Whanganui's headwaters were diverted and canalized by the Tongariro Power Development scheme (TPD), which "beheaded" (the term used by engineering planners at the time) some 90 streams and rivers across 320 square kilometres of watershed.[19] Rivers were tunnelled underground, around the mountain and north into Lake Taupō, to provide additional water to the eight dams on the Waikato River (New Zealand's longest). The TPD provides about 5 percent of the country's total electricity generation. For Māori of the River the scheme's effect on the flow, and so on their sustenance and spiritual life force, was profound. But the sunk costs – ecological damage to river health, and cultural damage, for Māori in particular – were not counted in public until the force of the Treaty was revived by those who appeared before the courts and the Waitangi Tribunal.

As the current dwindled, the tribe suffered.[20] But when designs were commissioned for the TPD in 1958, Whanganui tribal world views were not a factor. A key planner, reporting on a fiery 1964 public meeting about the scheme, observed the water levels would drop, but that change was of little long-term consequence:

> The value of such a change cannot be assessed in money; it is a matter of aesthetics and civic pride. The position can be alleviated by letting water spill into old channels, but this cannot be promised in advance and needs investigation into the economics of wasting water. As has been said before, after one generation, standards by which the matter will be judged will have changed in any case.[21]

Public opposition to the TPD did not delay its construction. When communities argue about dams, Veronica Strang suggests the outcome – "whether societies dam the flow or go with it – depend on who owns and controls 'living water' and therefore directs decision-making processes."[22] In "beheading" the Whanganui's headwaters, the power-broker's control was total and exclusive. So in the late 1980s, the Whanganui tribes took a case to an environmental planning tribunal, contesting the renewal of the state power company's right to take water for the TPD. The tribal focus was river traditions, and the negative ways the TPD affected both the tribes and the River itself. The government sought to keep the power flowing.

I recall the day when the evidence from that state Electricity Corporation was wheeled into dad's office: a ten-foot tower of dark red folders embossed with gold lettering, the fruit of much labour and many millions of dollars. I paled, but Martin gave an incredulous shrug, then got to work. The Māori evidence he helped present was less voluminous than Electricorp's, but it ran deep. River elders, the anthropologist Dame Joan Metge and psychologist Professor James Ritchie provided evidence on river tradition: the Whanganui elder Taitoko Tawhiri saw the diversion as akin to beheading a person or an eel; "the spiritual elements of the river itself, of its headwaters are being driven away."[23] Historians traced the almost total destruction of the eel weirs around the 1880s (for tourism and steamers). Over 350 weirs anchored life ways, and tribal hospitality (and thus *mana* or identity, status and standing). Energy analysts and economists detailed the problems with assuming a constant need for growth. There were many Pākehā allies in the hearings. Māori values were recognized in the Planning Tribunal's decision: some water was returned seasonally as a result of the case. But the compromise was insufficient for Māori who were frustrated with unresponsive laws. In 1990, Whanganui River tribes lodged a claim before the Waitangi Tribunal.[24] Dame Tariana Turia affirms:

> The spirit of the claim has never wavered. The people of Whanganui originally possessed and controlled the Whanganui river and all the things that gave the river essential life, and such possession and control has never been willingly relinquished. That is how it has always been. Whanganui have the longest litigation in the history of this country over their ancestral river – *te awa tupua*.[25]

The claim was urgent. As Sian Elias, Kathy Ertel and my father told the Tribunal, while the Crown keeps (re)learning, the people are dying.[26]

Oral tradition, Treaty interpretation, environmental and spiritual elements are at the heart of the Waitangi Tribunal's quasi-judicial inquiries, which are often held in tribal ancestral meeting houses, where elders set out, again and again, the kinds of remembering of river and community that counter the TPD planners' faith in hydro-amnesia. In his evidence before (yet another) Waitangi Tribunal inquiry, Ngāti Rangi's Keith Wood emphasizes the impact of the diversion, yet confirms the power of tribal memory:

> The imposition of the TPD [in 1974] and the diversion of water from our *tūpuna awa* [ancestral river] into other catchments is a deep grievance for Ngāti Rangi. We are not prepared to wait another generation and a half before our cultural and spiritual values would be addressed. While our rivers suffer, so do our people.
>
> Again, we remain resolute that our ancient values and practices are of critical relevance today in our holistic view of sustainable environmental management, given the extensive environmental degradation that has taken place globally over the last 150 years and global/national water shortages predicted to become some of humankind's greatest resource challenges in the next ten years. … We resolve to sustain our customary and ancestral relationships with our *tūpuna awa*, and as *tāngata tiaki* [guardians], to sustain our *mouri kōrero* [healing connections with life essence]. They are of critical relevance today and must be protected for the future. The Crown has been negligent in its duties to protect these *taonga* [treasures] not only for *tāngata whenua* but for all people in the community.[27]

Biocultural entanglement means Māori ritual and well-being pivots around water health, and the very continuities of flow. Remaining vigilant, the River tribes offer their traditions in the spirit of cooperation, displaying extraordinary generosity towards wider community well-being in an era of broken rivers.

The Tribunal's 1999 report on the Whanganui River claims recommended full ownership of the water and riverbed be returned to Māori. The government disagreed. In its later inquiry into the TPD, the Waitangi Tribunal found that a lack of Crown consultation and collaboration constituted a Treaty breach. As the Tribunal noted, when "use rights have become, de facto, property rights," Māori have little say.[28]

In 2012, several Māori tribes challenged the Crown's plan to privatize state power companies, in pursuit of Treaty claims to water ownership: "For Māori it becomes even more difficult to have *tino rangatiratanga* [full authority] recognised if resources pass out of Crown ownership."[29] During that litigation, Whanganui River tribes continued to negotiate a separate River settlement. The Whanganui tribes' agreement to prioritize control rather than ownership may have given them extra leverage in the final settlement negotiations, because the resulting legislation denies the concept of ownership in favor of giving the tribes more control over the kind of values and conversations decision-makers have, and making more equal conversations happen in the first place.

TE AWA TUPUA TREATY SETTLEMENT: ANCIENT FLOW, MODERN LAW

After 125 years of legal struggle, Whanganui River Māori have almost settled their Treaty claims in a Bill introduced to Parliament on May 2016 honouring *Te Awa Tupua*, the river of ancestral presence and power. The initial reading of the Bill received the unanimous support of every Member of Parliament, and speeches from the floor were full of praise, relief and celebration. The law records four (acclaimed and admired) paramount principles that I quote here:

1. *Ko te Awa te mātāpuna o te ora* (The River is the source of spiritual and physical sustenance)
   Te Awa Tupua is a spiritual and physical entity that supports and sustains both the life and natural resources within the Whanganui River and the health and well-being of the iwi, hapū and other communities of the River.
2. *E rere kau mai te Awa nui mai te Kahui Maunga ki Tangaroa* (The great River flows from the mountains to the sea)
   Te Awa Tupua is an indivisible and living whole from the mountains to the sea, incorporating the Whanganui River and all of its physical and metaphysical elements.

3. *Ko au te Awa, ko te Awa ko au* (I am the River and the River is me)
   The iwi and hapū of the Whanganui River have an inalienable interconnection with, and responsibility to, Te Awa Tupua and its health and well-being.
4. *Ngā manga iti, ngā manga nui e honohono kau ana, ka tupu hei Awa Tupua* (The small and large streams that flow into one another and form one River)
   Te Awa Tupua is a singular entity comprised of many elements and communities, working collaboratively to the common purpose of the health and well-being of Te Awa Tupua.[30]

Who gives voice to the River and ensures these principles remain paramount? Two appointees (one government, one Māori) monitor adherence to these principles and to river health, and are answerable first and foremost to the River. In bringing *mātauranga Māori*, Māori wisdom and tradition, into the Western legal arena in alluring new ways, the agreement draws on metaphor to convey the desire for ecological and cultural healing.

Whanganui *iwi* (tribes) are creating the opportunity for many to open to the spiritual landscapes their elders have safeguarded and shared. Those elders did not stint in their generosity of vision. Throughout the 125 years, there have been times of peaceful non-violent protest that drew in allies from all walks of life, a social movement alive to the Treaty's quest for coexistence, the double-hulled *waka* of *tangata whenua* (indigenous people of the land) and *tangata Tiriti* (the people of the Treaty who came after Māori). The inclusive, richly realized way their traditions inform this contemporary settlement is a generous *taonga*, a treasure, to the world. The River has always been ready to be heard. More people *are* listening to the River.

To formalize a genuine conversation, the work will be overseen by a standing committee of guardians, *Ngā Tāngata Tiaki o Whanganui*. That committee "will be recognised as having an interest greater than and separate from any interest in common with the public generally."[31] This triggers leverage within resource management decision-making. A collaborative process, inclusive of key persons and agencies with an interest in the River, will identify issues relating to the "environmental, social, cultural and economic wellbeing of *Te Awa Tupua*." That process takes its guiding metaphor for unity of purpose from the migration of the longfin eel (from Whanganui all the way to the warm Pacific waters near Tonga). The eel, long a customary and spiritual focal point for river life (as a delicacy, and a totem of the *kaitiaki* or spiritual guardians of rivers), is under threat from habitat loss, pollution and overfishing, but remains crucial in the exercise of biocultural connection.

REWEAVING THE PINIONED STREAMS

Alex Calder's 2012 work *The Settler's Plot: How Stories Take Place in New Zealand*, (arguably New Zealand's first ecocritical "settler studies" monograph), attests to activist poet James K. Baxter's legacy at the Whanganui River. His influence was borne of his commitment to the long-standing combined Māori and Catholic traditions on the Whanganui.[32] Calder describes Baxter as a "*Pākehā* pilgrim." Adept at "self-fashioning" and bridging cultures, he cited Māori as a source of healing, all the while largely romanticizing nature.[33] In *Jerusalem Daybook* (1971), Baxter critiqued the "soul numbing" characteristics of contemporary society that restricted people from acting communally, when collaborative values were ceasing to exist. His response was to identify five water-worn stones chosen from the Whanganui River, five spiritual aspects of Māori communal life which could be used to renew our freedom to cooperate with and relate to other human beings:

arohanui: the love of the Many;
manuhiritanga: hospitality to the guest and the stranger;
kōrero: speech that begets peace and understanding;
matewā: the night life of the soul;
mahi: work undertaken for communal love.[34]

These expressions will be central to the engagement necessary to bring the *Te Awa Tupua* settlement to life.

John Newton favors Baxter's "ingenuous" cross-cultural aspirations, as *Pākehā* are made to affirm an "on-going process of exploration, negotiation and critique" opening to the nature of *Pākehā* privilege in order to find a way beyond such exploitative inheritances, towards an awareness of white indebtedness to Māori.[35] The transmission of River gifts ripples outwards from the *Tira Hoe Waka*, transcending commercial imperative, community loss and indifference. Newton argues the language *Pākehā* use in the process of reconciliation must change, thereby enabling a longer conversation less constrained by either diffidence or caution. For Newton, "a vocabulary of 'plunder,' transgression and guilt needs to be expanded to encompass desire, debt and gratitude."[36] The focus shifts from shame to gratitude, particularly in the context of River tribes offering an inclusive view of river restoration as a mode of wider community restoration. The process will be pluralist, and will need to account for the varied ways *Pākehā* relate to place and spirituality in New Zealand. There are powerful Catholic and syncretic Whanganui River practices alive today, alongside a *Pākehā* world view tired of old orthodoxies and the limits of good intentions.[37]

As the *Te Awa Tupua* agreement beds in, it will demand agencies "recognise and provide for" *Te Awa Tupua* and its principles. The settlement requires a shift to relational thinking: it "changes the lens through which people – and particularly decision-makers – must view and relate to the river."[38] The people of the River, and the *mana* of the River itself, have already transformed both peoples' sense of the potential of water, identity and story. Yet as Maria Bargh observes, neoliberal values eschew the more collaborative values of *Pākehā* culture. There is much intra- and inter-cultural work to be done in an age when profit talks louder than the river; the resulting pollution, from farming and cities, is one legacy of an unsustainable model of economic growth. The Land and Water Forum, an independent collaborative group, has produced four reports (at government request) on water policy. In 2015, the Forum observed "the great bulk of our recommendations are yet to be implemented. … Implementation on the ground is urgent."[39] The *Te Awa Tupua* agreement recognizes the need for restoration on multiple levels. Yet overarching national policies are contested. In 2016, the government favours irrigation and development over stronger regulation of water quality. But diplomats of the world are invited to the signing of *Te Awa* Agreement. Many are pondering ways to accommodate (or perhaps contain) indigenous peoples' river relationships.

* * *

I am standing with Keith Wood of Ngāti Rangi, in a remote area as he observes the beautiful Wahianoa River tumble off the slopes of Mount Ruapehu. The water spangles across volcanic rock, only to be utterly swallowed by a TPD culvert. The beautiful words of the legislation have not replaced the eel's head. Yet tribal calls for restored fluvial connectivity means the power company has deigned to return a fraction of the water, through a small hosepipe connection at the culvert's edge. Some hope is evident in the trickling gesture, but its parsimony (the market rules) sometimes means the old despair is barely leavened. The work of reconciliation of waterways and communities is ongoing, but at least now the River has a stronger voice.

Rivers and healing linked in profound ways in the days before my father's death in 2003. His body pared back by a cancer's ineluctable passage, he was helicoptered to the headwaters of Te Awa Tupua and flown along its pulsing length. After the trip he attained a peace which helped him to transition to other worlds, to leave us, as I scattered drops of water from the River on his brow, his hands, his heart. Last sigh. Drops fell, and the waters rippled. When Whanganui Māori honored dad with a traditional farewell ceremony our family was embraced once more. The Treaty- and heart-driven seekers pass on the gift of healing, of endurance, of aroha.

* * *

At Martin's funeral, Professor Whatarangi Winiata, a leader in the area of Māori and Treaty rights, language revival, education, broadcasting, development and community, spoke. His 1983 invitation to my father drew him deep into the purpose and promise of the Treaty of Waitangi:

> A recurring theme in Martin's growing stature as a bearer of the Treaty was his insistence on the Treaty as a covenant, a sacred undertaking. His faith was central to this understanding. His journey into the complexities of the Treaty revealed an unswerving commitment to its underlying principles summarised for us by Māori Marsden:
> *Rangatiratanga* (chieftainship)
> *Whanaungatanga* (familial-community bonds)
> *Tohungatanga* (scholarship)
> *Manaakitanga* (generosity)
> *Ūkaipō* (belonging)
> *Kōtahitanga* (unity)
> … These values define Māori – the *tangata whenua* [people of the land, indigenous] partner to the covenant. These *kaupapa* [values, goals] sat easily on Martin's heart.[40]

He was sharing a pathway for others to follow. Treaty principles remain crucial, for – as Winiata shows – the Treaty's underlying principles are relational. They offer a form of ethical guidance and connection in intercultural dialogue born not of fear, but of the Treaty's potential for connection, its ability to fulfil a promise that yet endures.

CONCLUSION

Māori cultural scholar Irihapeti Ramsden explains how Māori modes of compassion encourage cross-cultural growth and the healing necessary for *all* peoples to flourish:

> It seems to me that reality can only be encountered through accepting the validity of the experiences of others. This means that the truths of others, the history of others, their justice and what it means to them to be free are real to them and are therefore legitimate in the world. The force which enables people to understand the legitimacy, the reality of others is *aroha*. *Aroha* therefore is of itself a force for freedom. Freedom of the human spirit. Freedom to be without fear.[41]

Ramsden writes of enacting "a cross-hatching of movement in and out of worlds," a movement characterized both by discomfort and homecoming.[42] This border work across cultures has been rewoven and gifted consistently by the River people, even when they felt silenced by the majority. They have found a way through with the power of memory, determination and *aroha*. As Charlotte Šunde notes, "The dual tensions that appear to divide may also be seen as creative tensions with potential to unite."[43] Tribal negotiator Brendan Te Tiwha Puketapu notes the emphasis for the tribe is on enabling relational thinking for the community: "We aren't here to talk about us. We're here to talk with you for the health of the Whanganui River."[44]

The elegance of the *Te Awa Tupua* agreement should not occlude the fact the Whanganui is polluted, burdened by land run-off, and it *remains* diverted. In 2016, the government is weakening environmental laws, and the matter of water ownership and corporate power is unresolved. These issues will not go away. Yet within the nascent unfurling of plural co-governance for the *Awa*, peoples' experiences on and with the River can create an opening for new ways of living in a more harmonious relationship with "nature" and its human communities. *Te Awa Tupua* is a new (and ancient) cultural story rooted in the sacred. It is part of an emerging global conversation and awareness that enables us to see ourselves and nature in terms of kinship, community, relationship and interconnectedness. It's a gift that is still unfolding.

Te Awa Tupua offers – and requires – holistic attention, restoration and honouring. I return to the capacity of water to heal, and for the best intentions of community to foster *aroha* for this wounded river and its stories. Change by the majority culture here is not impossible, the attentive have already realized the riches of biculturalism in language and shared experience. Again the Treaty of Waitangi can become a source for reconciliation. David Young reminds us of a phrase: "'The Whanganui is about the stuff that opens doors,' [River Māori] say. The doors are still opening."[45] A way is open for the legal system and the wider community to honour a form of embodied, storied collaboration. Measured against the intractable devastation writ large on waterways of this earth, there is reason for tentative hope (and for measured gratitude) that enough intercultural potential remains to help regift what this era of remnant demands. There is story and (some) biota still in those rivers yet tunnelled under and around mountains. Faith is vested in the waters again.

Many have journeyed with the river: it may be time to recognize it has claimed us. It has taken our hearts into its depths, left us on its banks, where water laps alongside stories, where the names of rapids and songs revive a past that's been flowing in plain sight. *Te Awa Tupua* has delivered Aotearoa New Zealand to a place where new possibilities stirred up by the journey might take hold.

Thanks to my family, Peter Horsley, Peter Wilson, Richard & Tom Eats, Keith Wood, David Young, the editors and an anonymous reviewer for their support or comments on earlier versions of this chapter. I am grateful for permission to reproduce portions of evidence from Keith Wood, Professor Winiata's eulogy, and some fragments from my previous article in Philosophy, Activism, Nature 4 (2007).

NOTES

1. Veronica Strang, *The Meaning of Water* (Oxford: Berg, 2004), 252.
2. Fifteen percent of New Zealand's 4.5 million people self-identified as Māori in 2015.
3. Waitangi Tribunal, *The Whanganui River Report* (Wai 167, 1999), 90, 91.
4. M. Dawson, et al., Submission on behalf of the International Commission of Jurists to the Waitangi Tribunal. In Waitangi Tribunal, *Report on the Te Reo Claim* (Wai 11, 1986). (Wellington, 1993 ed.), 21.
5. Michael Belgrave, "Looking Forward: Historians and the Waitangi Tribunal," *New Zealand Journal of History* 40:2 (2006), 230–50.
6. Alan Ward, *An Unsettled History: Treaty Claims in New Zealand Today* (Wellington: Bridget Williams Books, 1999), 167.
7. *New Zealand Maori Council v AG* 1 (1987) *New Zealand Law Reports* 64: 664.
8. Te Ahukaramū Charles Royal, "Tangaroa: The Sea. Water as the Source of Life." Online. *Te Ara – the Encyclopedia of New Zealand*, http://www.teara.govt.nz/en.
9. Linda Te Aho, "Ngā Whakataunga Waimāori. Freshwater Settlements," in *Treaty of Waitangi Settlements*, ed. Nicola Wheen and Janine Hayward (Wellington: Bridget Williams Books, 2012), 103.
10. *Ruruku Whakatupua Te Mana o te Awa Tupua* (Agreement between Whanganui River Iwi and the Crown), (2014), 16.
11. Hon. Tariana Turia, "Speech to Healing our Spirit: Worldwide Conference" (Kualoa Park, Honolulu, Hawaii, September 9, 2010), http://beehive.govt.nz/speech/healing-our-spirit-worldwide-conference-6th-gathering.
12. Karen Sinclair, *Prophetic Histories: The People of the Māramatanga* (Wellington: Bridget Williams Books, 2002), 195. Sinclair canvasses tribal syncretic practice in detail.
13. Donald Worster, *Rivers of Empire. Water and the America West* (New York: Pantheon, 1985), 325.
14. Ivan Illich, *H2O and the Waters of Forgetfulness: Reflections on the Historicity of "Stuff"* (Dallas: The Dallas Institute of Humanities and Culture, 1985), 24.
15. Joan Metge, *Tauira: Learning in the Māori World* (Auckland: Auckland University Press, 2015).
16. Jennifer Lawn, "Settler Society and Postcolonial Apologies in Australia and New Zealand," *Sites* 5:1 (2008), 34.
17. Keith Wood, "Waitangi Tribunal Evidence Doc #E30," *Te Kāhui Maunga: The National Park District Inquiry Report* (Wai 1130, 2013), 5.

18. David Young, *Rivers: New Zealand's Shared Legacy* (Auckland: Random House, 2014), 48.

19. Ministry of Works, *Tongariro Power Development: Background and History to Committal* (Wellington: Ministry of Works, 1981), 99, 124.

20. The online Waitangi Tribunal reports on river claims are replete with such examples.

21. Ministry of Works, "Taumarunui Borough Council Reaction 1964," in *Tongariro Power Development: Background and History to Committal* (Wellington: Ministry of Works, 1981), 135.

22. Veronica Strang, "Going against the Flow: The Biopolitics of Dams and Diversions," *Worldviews* 17 (2013), 171; Brahma Chellaney, *Water, Peace, and War: Confronting the Global Water Crisis* (Lanham: Rowman & Littlefield, 2013), 307.

23. M. Dawson and C. Wickliffe, Closing Submissions before the Planning Tribunal, Appeals 781/88 & 840/88 (Wellington: Luckie Hain, 1990) 40, 41.

24. Unlike other judicial bodies, the Tribunal is empowered to inquire into claims of Treaty breach, and measure claims against what the Treaty demands, as opposed to what is solely in statute. The report might (if the Crown so deigns) contribute to entirely separate negotiations with the government of the day for financial redress (a mere 1–2% of land value lost), return of Crown-owned land, cultural redress and relationship agreements with state agencies. See Maria Bargh, "The Post-settlement World (So Far): Impacts for Māori," in Wheen and Hayward, eds., 166–81 (*Op. Cit.* note 10), for some of the limitations of the Treaty settlement process.

25. Helen Leahy, *Crossing the Floor. The Story of Tariana Turia* (Wellington: Huia, 2015), 167.

26. Sian Elias, Martin Dawson and Kathy Ertel, Closing Submissions before the Waitangi Tribunal Whanganui River Inquiry, 1994. Wai 167, doc D18, 2. Subsequent counsel Jamie Ferguson reiterates the point.

27. Keith Wood, "Waitangi Tribunal Evidence Doc #E30" (Wai 1130), 15.

28. Waitangi Tribunal, *Te Kāhui Maunga: The National Park District Inquiry Report* (Wai 1130, 2013), 1163.

29. Maria Bargh, "Submission on Water Issues in Aotearoa New Zealand" (Geneva: UN, OHCHR, 2007), 14.

30. *Ruruku Whakatupua*, 7.

31. Electoral Commission, *Whanganui River Settlement. Ratification Booklet* (Wellington: Elections NZ, 2014), 28.

32. See Sinclair note 13.

33. Alex Calder, *The Settler's Plot: How Stories Take Place in New Zealand* (Auckland: Auckland University Press, 2011), 18, 22.

34. Cited in John Newton, *The Double Rainbow. James K. Baxter, Ngāti Hau and the Jerusalem Commune* (Wellington: Victoria University Press, 2011), 11.

35. John Newton, "Becoming Pākehā," *Landfall* 218 (2009), 39.

36. Ibid.

37. Patricia Laing, "Spirituality, Belief and Knowledge: Some Reflections on Constructions of Maori Health," in *Plural Medicine*, ed. Waltraud Ernst (Routledge: London, 2002), 153–70; Sinclair, *Prophetic Histories*; Strang, "Going Against," 94–96.

38. Electoral Commission, *Whanganui River Settlement*, 25.

39. Land and Water Forum, *Fourth Report of the Land and Water Forum* (Wellington, 2015), vii, 1.

40. Whatarangi Winiata, "Eulogy for Martin Dawson" (May 19, 2003). See also Māori Marsden, *The Woven Universe: Selected Writings of Rev. Māori Marsden*, ed. Te Ahukaramū Charles Royal (Ōtaki: Estate of Rev. Māori Marsden, 2003) for more on these concepts.

41. Irihapeti Ramsden, "Borders and Frontiers," in *Regaining Aotearoa: Maori Writers Speak Out*, ed. Witi Ihimaera (Auckland: Reed, 1993), 345.

42. Ibid., 344.

43. Charlotte Šunde, "The Water or the Wave? Toward an Ecosystem Approach for Cross-Cultural Dialogue on the Whanganui River, New Zealand," in *Complexity, Uncertainty, and Managing for Sustainability: The Ecosystem Approach*, ed. David Waltner-Toews, James J. Kay, and Nina-Marie E. Lister (New York: Columbia University Press, 2012), 248.

44. Young. *Rivers*, 42.

45. David Young, *Woven by Water: Histories of the Whanganui River* (Wellington: Huia, 1998), 177.

BIBLIOGRAPHY

Bargh, Maria. "Submission on Water Issues in Aotearoa New Zealand." Geneva: UN, OHCHR, 2007.

———. "The Post-Settlement World (So Far): Impacts for Māori." *Treaty of Waitangi Settlements*. Ed. Nicola Wheen and Janine Hayward. Wellington: Bridget Williams Books, 2012, 166–81.

Belgrave, Michael. "Looking Forward: Historians and the Waitangi Tribunal." *New Zealand Journal of History* 40:2 (2006), 230–50.

Calder, Alex. *The Settler's Plot: How Stories Take Place in New Zealand.* Auckland: Auckland University Press, 2011.

Chellaney, Brahma. *Water, Peace, and War: Confronting the Global Water Crisis.* Lanham: Rowman & Littlefield, 2013.

Dawson, Martin and C. Wickliffe. "Closing Submissions before the Planning Tribunal, Appeals 781/88 & 840/88." Wellington: Luckie Hain, 1990.

Dawson, Martin et al. "Submission on Behalf of the International Commission of Jurists to the Waitangi Tribunal." *Report on the Te Reo Claim, Wai 11.* Ed. Waitangi Tribunal, Wellington, 1993.

Electoral Commission. *Whanganui River Settlement. Ratification Booklet.* Wellington: Elections NZ, 2014.

Elias, Sian, Martin Dawson and Kathy Ertel. "Closing Submissions before the Waitangi Tribunal Whanganui River Inquiry, Wai 167, doc D18." (1994).

Illich, Ivan. *H2O and the Waters of Forgetfulness: Reflections on the Historicity of "Stuff."* Dallas: The Dallas Institute of Humanities and Culture, 1985.

Laing, Patricia. "Spirituality, Belief and Knowledge: Some Reflections on Constructions of Maori Health." *Plural Medicine.* Ed. Waltraud Ernst. Routledge: London, 2002, 153–70.

Land and Water Forum. *Fourth Report of the Land and Water Forum.* Wellington, 2015.

Lawn, Jennifer. "Settler Society and Postcolonial Apologies in Australia and New Zealand." *Sites* 5:1 (2008), 20–40.

Leahy, Helen. *Crossing the Floor. The Story of Tariana Turia.* Wellington: Huia, 2015.

Marsden, Māori. *The Woven Universe: Selected Writings of Rev. Māori Marsden.* Ed. Te Ahukaramū Charles Royal. Ōtaki: Estate of Rev. Māori Marsden, 2003.

Metge, Joan. *Tauira: Learning in the Māori World.* Auckland: Auckland University Press, 2015.

Ministry of Works. "Taumarunui Borough Council Reaction 1964." *Tongariro Power Development: Background and History to Committal.* Wellington: Ministry of Works, 1981.

Ministry of Works. *Tongariro Power Development: Background and History to Committal.* Wellington: Ministry of Works, 1981.

New Zealand Maori Council v AG 1 (1987) *New Zealand Law Reports* 64.

Newton, John. "Becoming Pākehā." *Landfall* 218 (2009), 38–48.

———. *The Double Rainbow. James K. Baxter, Ngāti Hau and the Jerusalem Commune.* Wellington: Victoria University Press, 2011.

Ramsden, Irihapeti. "Borders and Frontiers." *Regaining Aotearoa: Maori Writers Speak Out.* Ed. Witi Ihimaera. Auckland: Reed, 1993.

Royal, Te Ahukaramū Charles. "Tangaroa: The Sea. Water as the Source of Life." Online. *Te Ara – the Encyclopedia of New Zealand.* http://www.teara.govt.nz/en.

Ruruku Whakatupua Te Mana o te Awa Tupua (Agreement between Whanganui River Iwi and the Crown), (2014).

Sinclair, Karen. *Prophetic Histories: The People of the Māramatanga.* Wellington: Bridget Williams Books, 2002.

Strang, Veronica. *The Meaning of Water.* Oxford: Berg, 2004.

———. "Going against the Flow: The Biopolitics of Dams and Diversion." *Worldviews* 17 (2013), 161–73.

Šunde, Charlotte. "The Water or the Wave? Toward an Ecosystem Approach for Cross-Cultural Dialogue on the Whanganui River, New Zealand." *Complexity, Uncertainty, and Managing for Sustainability: The Ecosystem Approach.* Ed. David Waltner-Toews, James J. Kay, and Nina-Marie E. Lister. New York: Columbia University Press, 2012.

Te Aho, Linda. "Ngā Whakataunga Waimāori. Freshwater Settlements." *Treaty of Waitangi Settlements.* Ed. Nicola Wheen and Janine Hayward. Wellington: Bridget Williams Books, 2012.

Turia, Hon. Tariana. "Speech to Healing our Spirit: Worldwide Conference" (Kualoa Park, Honolulu, Hawaii, September 9, 2010). http://beehive.govt.nz/speech/healing-our-spirit-worldwide-conference-6th-gathering.

Waitangi Tribunal. *The Whanganui River Report.* Wai 167, 1999.

———. *Te Kāhui Maunga: The National Park District Inquiry Report,* Wai 1130, 2013.

Ward, Alan. *An Unsettled History: Treaty Claims in New Zealand Today.* Wellington: Bridget Williams Books, 1999.

Winiata, Whatarangi. "Eulogy for Martin Dawson" (May 19, 2003).
Wood, Keith. "Waitangi Tribunal Evidence Doc #E30." *Te Kāhui Maunga: The National Park District Inquiry Report*, Wai 1130, 2013.
Worster, Donald. *Rivers of Empire. Water and the America West*. New York: Pantheon, 1985.
Young, David. *Woven by Water: Histories of the Whanganui River*. Wellington: Huia, 1998.
———. *Rivers: New Zealand's Shared Legacy*. Auckland: Random House, 2014.

Chapter 4

From Wai 262 to Water

Towards Postcolonial Property Rights in Aotearoa New Zealand

Jacob Otter

In 2012, Aotearoa New Zealand's (NZ's) conservative-led government began to roll out its long-threatened partial privatization of many of the nation's state-owned enterprises (SOEs). Power companies were among the first SOEs on the fiscal chopping block and included Mighty River Power, a power company that generates electricity from geothermal sources in the Taupō region, and Aotearoa New Zealand's longest river, the Waikato River. Despite claims that the election gave them a mandate to go ahead with the sell-off, the government's plans were extremely controversial and met with noisy resistance. This resistance coalesced into two largely unrelated strategies. The first was a petition lead by the Green Party that gained enough signatures to force a referendum on the government's plans. A total of 1.4 million votes were then cast on the question, "Do you support the Government selling up to 49% of Meridian Energy, Mighty River Power, Genesis Power, Solid Energy and Air New Zealand?" Two-thirds of the votes responded with a "No" but the National party government was not to be swayed, and ignored the referendum.

The second strategy began with the filing of two claims to the Waitangi Tribunal by the NZ Maori Council (NZMC). The Waitangi Tribunal is Aotearoa New Zealand's pre-eminent forum for official inquiries into Crown breaches of the Treaty of Waitangi, meaning that the claims by the NZMC were to be a lot harder for the government to ignore.[1] The claims asserted that the partial privatization of the power companies would breech Māori customary ownership rights of fresh and thermal waters, and impede the Crown's ability to compensate Māori for historical breeches of the Treaty of Waitangi. In their submission to the Tribunal, the claimants summarized the two claims as follows:

> 2358 – *that Māori have proprietary rights to freshwater and geothermal resources, the protection of which was guaranteed under Article 2 of the Treaty,*

> and that the Crown has breached that guarantee of protection by refusing to recognise those rights and by establishing and maintaining a management regime that prevented Māori owners from exercising the full enjoyment of those rights; and
>
> 2357 – that the Claimants have unresolved claims in relation to freshwater and geothermal resources including Wai 2358 and including historical claims lodged by each of the co-claimants, that the Crown intends a partial sale of its power-generating SOE assets, and by inhibiting resolution of these outstanding claims the sale will cause prejudice to Māori in breach of the Treaty.[2]

The key point in the summary above, and what would become the focus of the Crown's counter-argument, is "that Māori have proprietary rights." If the Crown could find a way to counter this point there was every chance the Tribunal might give them favourable findings, and they sought to make this point by using the findings of one of the most significant claims to come before the Waitangi Tribunal, Wai 262.

In this chapter, I examine the Crown's use of the Wai 262 against Māori assertions of their rights to freshwater as an entry point onto a broader set of questions about the limits to non-indigenous conceptions of indigenous ontologies, due to the central role the nature-culture distinction plays in settler modernity. The Crown's use of Wai 262 findings as a shield to block Māori assertions of customary rights to freshwater raises serious questions as to how the Waitangi Tribunal interpreted the Wai 262 claim, especially the claim's key assertions as to the nature and vitality of taonga.[3] I begin by exploring the political ontology of *taonga*, the way in which it operates to generate connection and entanglements between people, places and things in *te ao Māori* (the Māori world). However, my exploration of this entanglement is not just a mapping exercise – rather I seek to emphasize the inherently political dimensions of Māori relationships between people and nonhuman entities. It is this political dimension that comes to the fore in the Wai 262 claim, whereby, as Amiria Henare argues, taonga come to encompass non-Māori "things" into relationships expressed on and in Māori terms. The Tribunal's response to this irruption of *taonga* was to find that Māori did not, indeed could not, have property rights in *taonga* as property is a concept that arises from *te ao Pākehā* (the dominant majority in Aotearoa New Zealand), whereas taonga are from *te ao Māori*. This finding contradicts the dynamic and encompassing traits of taonga that instigated the Wai 262 claim. I undertake an analysis of the Tribunal's report on Wai 262, *Ko Aotearoa Tēnei*, to explore how taonga are shifted into the partitioned, bicultural realm of *te ao Māori* and *te ao Pākehā*. As I show, this shifting of *taonga* brings us to one of the founding conceits of settler modernity, the nature/culture distinction, and along with it the violence necessary to force indigenous worlds into this partitioned Universe.

THE WAI 262 CLAIM AND KO AOTEAROA TĒNEI

The Waitangi Tribunal's findings on the Wai 262 claim were published in 2011, in a two-volume report called *Ko Aotearoa Tēnei*. Claimants from a number of iwi[4] first lodged the Wai 262 claim in 1991.[5] Considered by some as "the grandfather of treaty claims,"[6] Wai 262 had an extremely broad reach, and while it garnered significant global acclaim,[7] in Aotearoa New Zealand the Pākehā majority largely considered it a threat to "business as usual."[8] What made the Wai 262 claim so powerful and unique was the way in which it enabled and mobilized *taonga*. *Taonga* are important entities within *te ao Māori*. Typically translated as "treasure," *taonga* can include a range of human, nonhuman and more-than-human entities, and blurs distinctions between the material, the immaterial and the conceptual. The key feature of the Wai 262 claim was that Māori had property rights to these *taonga*, and that the Crown had failed its obligation to recognize and support the property right, such that many *taonga* were extinct or heading that way. However, when the Waitangi Tribunal to report on this issue in *Ko Aotearoa Tēnei*, they proposed a large number of programmes, policies and regulatory regimes to overturn the current discriminatory regime,[9] but to great surprise and controversy,[10] *Ko Aotearoa Tēnei* also stated that Māori do not have property rights to *taonga*.[11] The reason Māori do not have ownership rights in water is the property rights are of *te ao Pākehā* and are thus antithetical to *te ao Māori*. The Tribunal argued instead for a new category of rights that they called kaitiakitanga rights. I discuss this term *kaitiakitanga* below, but for now I would note that the Tribunal considered the codification of this term into a right to be a more accurate way for Māori to receive recognition of their rights to *taonga* than property rights.

However, for all the hard work and thought that the Tribunal put into how kaitiakitanga rights might be created and enforced, the ideas has not gained any political traction. Beyond a press release[12] and a media quote[13] the government has said very little about *Ko Aotearoa Tēnei* since the report's release, and an official response continues to seem unlikely. The reason for this is indicative of the wider lack of sympathy for Māori rights among the *Pākehā* majority of Aotearoa New Zealand's population.[14] However, as I noted above, the one place in which the Crown has used this distinction between property rights and kaitiakitanga rights, was before the Waitangi Tribunal to "shield" the Crown from the water claims:

> The Crown's second argument was that English-style ownership is not in fact the best English cultural equivalent for Maori rights. For this argument, it relied primarily on the Waitangi Tribunal's report on the "indigenous flora and fauna and Maori cultural and intellectual property claim" (Wai 262). In light of that report, the Crown suggested that kaitiakitanga is the true and practical

> expression of Maori rights in respect of environmental matters, including water resources. Whether it be full kaitiaki control, partnership (co-governance or co-management), or a lesser interest (kaitiaki influence through consultation), this is the correct mechanism to give expression to the rangatiratanga protected by the Treaty. Ownership is not the appropriate mechanism; as Justice Williams commented in the Wai 262 hearings, "there's no Maori word for ownership."[15]

As such, the Crown continued, they were more than happy to sit down with Māori and talk about kaitiakitanga rights to water, and happily pointed out that they were already doing so through the Iwi Leaders Group[16] and the Fresh Start for Fresh Water programme.[17] To further underline this point while also asserting their superior morality, the Crown also claimed that their management of water in this way is consistent with water being a resource of the commons. The Crown's role was to manage water allocation and use, as well as balance and reconcile the "multi-dimensional" interests that have a stake in it.[18] Such a balance includes, ironically enough, the right to instigate partial privatization of the power companies that otherwise rely on this commons.

This use of the property rights/kaitiakitanga rights distinction by the Crown was not just the result of some clever legal rhetoric but points to some fundamental limitations in *Ko Aotearoa Tēnei* that were open to exploitation. The limitations inherent to the property rights/kaitiakitanga rights distinction, can be contrasted to the entities that mobilized the claim, *taonga*. *Taonga* are agentic and vital, and play an important political role in negotiations between worlds. Later, I examine how *taonga* are represented in *Ko Aotearoa Tēnei* and document how Tribunal's process of analysis slowly transforms *taonga* from the agential entities in the Wai 262 claim into mute objects, amendable to use by the Crown.

TAONGA

The entities that are central to the Wai 262 claim are *taonga*. Jade Tangiahua Baker describes *taonga* as "objects once associated with ancestors continue to exude properties of ancestral efficacy. And although it is often demonstrated the role of people as determining the placement and position of an object, for Maori of Aotearoa New Zealand this role is fluid, and sometimes the object itself is the border negotiator."[19] I show here that *taonga* are dynamic and expansive, encompassing entities, people and processes. They are significant nodes in the political ontology of *te ao Māori*. Mario Blaser describes political ontology as the politics of practices that shape a world or ontology, and includes the conflicts and negotiations that emerge as worlds "strive to sustain their own existence as they interact and mingle with each other."[20] The practices that enable *taonga* are what the Wai 262 claimants argue have been endangered by the activities of the Crown. But as I will show, the Wai 262 claim itself becomes a *taonga*. Wai 262 forces domains such as politics and law to stretch and fit to include *taonga*, while the practices of negotiations and obligations that make space for *taonga*, transform these domains into *taonga* themselves. Many of the Tribunal's recommendations were aimed precisely at these kinds of transformations though none have been instituted. The diplomatic potential of *taonga* was thus lost amidst the unfolding of the Waitangi Tribunal's processes responding to the Wai 262 claim.

Taonga find their way into the Wai 262 claim through reference to them in Article Two of *Te Tiriti ō Waitangi*. The key section of Article Two states: "The Queen of England agrees to protect the chiefs, the subtribes and all the people of New Zealand in the unqualified exercise of their chieftainship over their lands, villages and all their treasures."[21] As I noted, *taonga* can include a wide array of entities, places, peoples, thing and even concepts. Paul Tapsell (Te Arawa) has written a lucid and considered account of *taonga* that sets out some of the term's broader connotations. Tapsell writes: "For Maori, if an item, object or thing is described as *he taonga* it immediately elicits a strong emotional response based upon ancestral experiences, settings, and circumstances. The underlying force driving this response to *taonga* is *whakapapa*."[22] The word *whakapapa* is commonly translated as "genealogy." Along with strong emotional responses, the quote from Baker that opened this section shows the imbrication of *taonga* with *whakapapa* means they are also fundamentally about politics, negotiation and power. As Tapsell identifies, at one level the connection between *whakapapa* and *taonga* "[reinforces] a kin group's complex identity and authority over their estates." But, he continues, *taonga* are more than "simple identity markers" within the webs of *whakapapa*, they also have their own force, power, and agency.[23]

Tapsell discusses three features of the agency of *taonga*. The first feature is *mana*, or authority and power. Tapsell argues that *mana* is an index of the ancestral effect of *taonga* via "direct association with ancestors as [the *taonga*] passes down the generations"; the greater the ancestor, the more *mana* the *taonga* has.[24] Along with *mana* comes the "complementary presence of *tapu*." *Tapu* is the "social controlling agent" of *taonga*, managing its relationships with other entities, including people.[25] Tapsell describes this management as *kaitiakitanga*. This is a trustee-like relationship, whereby senior elders and their families ensures that *taonga* are related to appropriately but also that the *taonga* is performed and made available to the kingroup in appropriate ways. The way in which these *kaitiakitanga* relations are enabled by the senior elders and their families is through their ability to recite the *korero*, or oratory and knowledge, particular to the *taonga*.

For Tapsell, *korero* is the most important element "contained within *taonga*": "Without *korero*, a *taonga* ceases to be recognised as representing a specific genealogical position for its descendants" and thus undermining the *kaitiakitanga* relationship, as well as the *mana* and *tapu* that inheres in the *taonga*.[26]

Tapsell describes the modes of agency that *taonga* have through two metaphors, the flight of the tui – a native New Zealand bird – and the orbit of the comet. The tui's flight path is "gliding and dipping … with elegant swooping motions" and has links this to the ancient meaning of tui, "stitch."[27] For Tapsell, the agency of *taonga* is like the flight of a tui, as the shift between movement and stillness is analogous to the appearance and reappearance of *taonga* at key events of a kin group.[28] This weaving quality of *taonga* connects past to present, and provides guidance and connection necessary for the continuity and well-being of the group. The orbit of the comet describes the agency of *taonga* that may move beyond the direct sphere and influence of the tribal group. These *taonga* are embodiments of important ancestor's authority, so their release by a tribe is especially significant. With the gifting of such *taonga*, Tapsell notes an increase in reciprocal expectation called *utu*, "a transfer of indebtedness."[29] Upon entering the orbit of another tribe, *taonga* "[continue] to carry the *mana*, *tapu* and *korero* of their former kin group, [and] now [acquire] a new layer of *korero* as it [becomes] woven into the *korero* of another tribe." These *taonga* may then "if a future occasion warranted it" be "gifted to yet another tribe" and it may be some time before the original tribe see that *taonga* again.[30] For, as *taonga* are the material embodiment and enactment of *mana*, *tapu*, and *korero*, they can hold a relationship between tribal groups where the indebtedness between them enables the beginning of a negotiation.

Utu is perhaps among the most renowned characteristics of *taonga* as it informed Marcel Mauss's famous depiction of the *hau* or "spirit of the gift." Henare notes that Mauss describes *taonga* as "the vehicle for its *mana*, its magical, religious, and spiritual force" and it is this force that contains "the obligation to reciprocate."[31] But whereas for Mauss *mana* retains an anthropocentricism, that is "Even when [a *taonga*] has been abandoned by the giver, it still possesses something *of* him,"[32] Henare shows how Mauss's source, Tamati Ranapiri, states that the taonga *is* the *hau* or spirit. That is, there is "a precise identity … between thing and spirit."[33] Because of this "precise identity," Henare contends that *taonga* are a "distinctively Māori register of value," a value whose *mana*, *tapu* and *korero* is recognized, staged, reasserted and negotiated with and among *kaitiaki*.[34] What is important about the Wai 262 claim then, and here I follow Henare, is the way in which it shows how *taonga*, as both concept and entity, is able to expand the way it registers value to encompass new forms.

That taonga are expansive, encompassing and generative is perhaps most forcefully and comprehensively enunciated in the Wai 262 Second Amended Statement of Claim from the claimants at Ngāti Kuri, Te Rarawa and Ngāti Wai:

> The [Wai 262] claim relates to te tino rangatiratanga of Ngati Kuri, Te Rarawa and Ngati Wai in respect of indigenous flora and fauna me o ratou taonga katoa (and all their treasures) within their respective tribal rohe, including but not limited to te reo, matauranga, knowledge systems, laws, customs and values, whakairo, waahi tapu, biodiversity, natural resources, genetics and genetics derivatives, Mäori symbols, images, designs, and their use and development and associated indigenous, cultural and customary heritage rights (including intellectual property and property rights) in relation to such taonga. "Taonga" in this claim refers to all elements of the claimants' estates, both material and non-material, tangible and intangible.[35]

It is visible in this statement that a vast range of objects, tangible and intangible, Māori and "western," are brought "into the sphere of taonga, encompassing them under the heading *me o ratou taonga katoa*" from Article Two of the Treaty of Waitangi.[36] The political ontology of *taonga* therefore is encapsulated in the way Wai 262 emerges from and extends the entanglements of people, places and things without drawing boundaries or demarcations around what *taonga* might be or might become.

TAONGA, KO AOTEAROA TĒNEI AND THE STATEMENT OF ISSUES

If the Wai 262 claim does not seek to draw boundaries between *taonga*, it is thus striking that one of the key findings of *Ko Aotearoa Tēnei* is that such a boundary is fundamental, namely the boundary between kaitiakitanga rights and property rights. It is this dichotomy that is used by the Crown against Māori as a "shield" to deflect Māori claims to property rights in freshwater. The Crown was able to insist that *taonga* and property were two different entities from two different worlds. To understand the emergence of this dichotomy and the emaciated version of *taonga* it presents, I seek to stay with *taonga* by following their transformation through the work of the Waitangi Tribunal in the lead up to the publication of *Ko Aotearoa Tēnei*. I focus on one key document in particular, the Statement of Issues (SoI). The transformation of *taonga* from agential and vital is a significant example of the silent antagonism, due to the way in which the more-than-human dimensions of *te ao Māori* are repressed. This repression is visible in the removal of the agential dimensions and the insistence that they are an epistemology which has significant consequences for the way in which law is able to recognize Māori worlds. It also brings to the fore one of the fundamental conceits of notions of property and how one may create it.

The SoI is a document that the Waitangi Tribunal develops early in the reporting process. The procedures that generate the SoI are known as "The New Approach," and these were developed in 2001–2 as a means to streamline and speed up the Tribunal's reporting process.[37] The aim of the SoI is to identify and distil the key aspects of the claim and provide the Crown's initial response. The SoI emerges after a sustained period of clarification, debate and research, and provides a template for the final report. The Issues are agreed upon by both parties and provide a shared understanding of what is at stake in the claim and what the Tribunal's subsequent research should focus on. The SoI can be considered the first iteration in a process of translation that makes the claim amendable to a governmental apology and compensation. The SoI for the Wai 262 claim was finalized in July 2006, and it does two important things: it sets out a definition of *kaitiaki* that will form the basis for kaitiakitanga rights, and it places *taonga* into four Parts or categories:

Part One: Intellectual Property Aspects of Taonga Works
Part Two: Biological and Genetic Resources of Indigenous and/or Taonga Species
Part Three: Tikanga Maori, Matauranga Maori and Te Reo Maori
Part Four: Relationship of Kaitiaki with the Environment

Each Part includes the Treaty breeches related to these taonga and Crown responses.

The reason for the Tribunal's redirection of taonga into these categories is so that they could "rationalise" the inquiry. The Tribunal discuss how the period just before the finalizing of the SoI was a very challenging period in the assessment of the claim as there was a lot of disagreement between the claimants, the Crown and the Tribunal as to the scope of the inquiry and how the issues were to be defined. In particular, the claimants wanted the historical and constitutional aspects of the claim considered, while the Crown did not want any replication or discussion of issues that they considered suited to a district inquiry[38]:

> In response to these differences, the Tribunal attempted to *rationalise* the inquiry process, initially in 1997 by proposing to devise a "schedule of issues" and subsequently in 2001 by making a decision to develop a "statement of issues" summarising the claims and providing the key questions on which the Tribunal would hear evidence.[39]

If the redirection of taonga into an example of rationality, what understanding is informing this rationality?

Linda Tuhiwai Smith has argued that "one of the supposed characteristics of primitive peoples was that we could not use our minds or intellects," so it would be a bold claim from the Tribunal to be seeking "rationality" amidst the knowledge practices of Māori.[40] Rather, rationality has links to Enlightenment discourse and the rise of modernity, along with a coeval relationship to colonization. The role of rationality is to insert the nature/culture dualism into the Wai 262 claim and redistribute *taonga* either side of this partition. This is evidenced by looking more closely at the Parts that make up the SoI. Part One becomes concerned with *taonga* that are Culture, Part Two with *taonga* that are Nature and Part Three is also about Culture. Part Four continues to work across the divide, although even here there is a sense of a division emerging with "Environment" anchoring the phrase and becoming a backdrop upon which *kaitiakitanga* takes place. The agentic dimensions to *taonga* are here pared back, and they are placed into what are considered to be the two defining realms of all peoples, Nature and Culture. Thus, *taonga* are rationalized by being redirected into the original dualism of Modernity and it is on this basis that they start to become visible and the Tribunal's hearings can proceed.

There is however a problem with the SoI's rationalization of *taonga* into the categories of Nature and Culture. This is that the traits that have historically distinguished *te ao Māori* and *te ao Pakeha* – the former is aligned with Nature, while the latter with Culture – also begin to collapse and any real difference gives way. Although this distinction could be seen as rather specious to begin with,[41] the Tribunal's own hearings and evidence from many Māori drew distinctions between western notions of property and "Māori control of Māori things." The binary distinction the Tribunal deployed to understand the difference between *te ao Māori* and *te ao Pākehā* needed to be an alternative framework to the Nature/Culture dualism as this was now being used as the foundation for both world views. This is a significant shift from the view of Māori as natural and primitive; the Tribunal gives both world views equal status as having acceptable Cultural interpretations of Nature. The alternative framework that would provide the foundation for this distinction between equals is the kaitiakitanga rights/property rights dichotomy. Kaitiakitanga rights are the appropriate basis for *te ao Māori*, whereas property rights are the appropriate basis for *te ao Pākehā*. But what the Tribunal does not seem aware of is that property rights are not just a Cultural feature of *te ao Pākehā*; they are a sanction of an ontological performance that is inherently colonial and only available to particular modes of interaction with Nature.

NATURE, COLONIALISM, PROPERTY

According to dominant understandings, the creation of a property right occurs by moving an entity from Nature to Culture via labour. John Locke's labour theory of appropriation is the most prominent example of this view, and has been heavily implicated in the justification of colonization.[42] This labour has had a number of names, but I want to go with the term "cultivation"; the etymology of the word culture has its root in cultivation.[43] It should be stated here that interactions with nonhuman life occur among all people all of the time. General terms for entities that emerge from the interface of human–nonhuman infra-action include hybrids, quasi-objects and nature-cultures. However, according to the Modern view popularized by Locke, only those who are Modern can apply Culture, while the designation of indigenous peoples as "primitive" and thus a part of Nature, means they are deemed unable to labour on Nature in a way able to engender property. A property right is thus the recognition of the "right kind" of Cultural interaction with Nature, or rather, it is a sanction for one specific mode of creating hybrid entities, a mode that can only be generated that generated by the Moderns and typically considered to be based on Science.[44] Non-modern modes of creating relationships of ownership and control of hybrids, such as that which occurs between *kaitiakitanga* and *taonga*, are not considered to create property. This issue of who gets to create property was what Māori were challenging with the Wai 262 claim, while much of the resistance to the claim was based on the belief that Māori modes of interaction with Nature were not considered the right kind of Culture to justify a property right.

The Waitangi Tribunal do not consider this ontological dimension to western property, thus leaving its colonial underpinning intact. Rather, the Tribunal get caught up in a bind by trying to establish a Cultural equality between kaitiakitanga rights and property rights. It is this bind that leads to the Tribunal position that Māori do have property rights. By describing *kaitiakitanga* relationships with *taonga* as equivalent to the nature/culture distinction, the Tribunal were forced to preserve the difference between *te ao Māori* and *te ao Pākehā* via the kaitiaki/property distinction. But in accepting that property rights are only available to *te ao Pākehā*, the Tribunal had to admit that Māori modes of creating hybrids – *kaitiakitanga* – do not amount to a property right. *Kaitiakitanga* is not considered to have the ability to sanction nature-culture relationships and is thus of a different quality to that of western property rights. But what is this quality?

In trying to uphold the cultural difference of Māori without suggesting they make property rights, the Tribunal must define the interactions that are specific to Māori in creating "taonga species" and "taonga works." As I have shown, only *te ao Pakeha* are considered to access Nature, because only its adherents do the right kind of labour that can achieve the right kind of sanction that shows they gain this access. The particular modes by which *taonga* mediate and negotiate between worlds, what Henare describes as the "object-like qualities of taonga," are disregarded by the Tribunal. Rather, the Tribunal, bound to the dictates of the Nature/Culture dualism, must find that Māori relationships with Nature do *not* access the world but are considered a particular kind of knowledge *about* the world. As the object-like qualities of *taonga* retreat from view, the Tribunal can only assert that *taonga* must exist in the mind, as defined by *whakapapa*. *Whakapapa* becomes a kind of meta-concept, standing above the actual politics and negotiations that take place amidst *taonga* and *kaitiaki*. They draw a term directly from western philosophy to describe this when they state that *whakapapa* is "not just a way of ordering humans and the world; it is an epistemology, a way of ordering knowledge."[45] This contrasts with the views of Tapsell and Tangiahua Baker, where *whakapapa* seem to flow from the relations established by *taonga*. Indeed, this is also consistent with the Wai 262 claim, where *whakapapa* extends and encompasses new entities, concepts and things.

At this point, I think we have come a very long way from the Statement of Claim discussed at the beginning of this chapter. The Wai 262 Claim was a statement about what exists and what is real. It was the enactment of an ontology. By the time we reach the SoI, these entities are being "re-raised" and "re-directed." And by the time we get to *Ko Aotearoa Tēnei* are no longer real and are instead the outcome of a theory of knowledge; they are reduced to a belief. Thus, the Tribunal's decision to not find property rights in kaitiakitanga relationships is not just a legal failure, it is also a failure to recognize the vitality and dynamicism of Māori worlds and the entities that populate it. The other side to all this is that it also leaves the particular hubris of *te ao Pakeha* in place.

The particular hubris of *te ao Pakeha* is that only they can access nature, only they can access nonhumans. It is this access to nature that makes *te ao Pakeha* superior, and even, beyond culture, connecting them to the broader hegemony of the West. As Bruno Latour puts it:

> We Westerners cannot be one culture among others, since we mobilise Nature. We do not mobilize and image or symbolic representation of Nature, the way other societies do, but Nature as it is, or at least as it is known by the sciences – which remain in the background, unstudied, unstudiable, miraculously conflated with Nature itself.[46]

What the application of property rights does then is authorize the *te ao Pakeha* mobilization of Nature, via Science, as the only correct way to know and access nonhumans. By not authorizing *kaitiaki* relationships as creating property rights, this relationship is considered less valid than Science as a way for knowing nonhumans. It also infers that the entities that populate *te ao Māori* are not considered to be real. Rather, these are an epistemology, a narrative, a story, certainly one that the Tribunal argues we should all be supporting and working to protect but ultimately, with a healthy dose of "reality," we know this is not how things really are. From the perspective of the tribunal, to know nature, to know nonhumans, we have to look to the West and to Science: this is why it makes sense to talk of "Māori culture" but not "Māori nature," for Nature belongs to science and to the west.[47]

CONCLUSION

The Wai 262 claim was a significant intervention into the colonial order of things, at once challenging dominant conceptions of how property can be made, while providing a significant opportunity for negotiation between worlds. The Tribunal, by failing to find that taonga could be property, did not rise to the challenge offered by the claim. Rather, by reinstituting rational thinking via the nature/culture divide *taonga* were stripped of their power to straddle and support negotiation between worlds. When the Crown were confronted by taonga in the Wai 2358 claim, this time in the guise of water, they were able to claim *Ko Aotearoa Tēnei's* emaciated version of *taonga* as the correct grounds for understanding the nature of Māori ownership rights to water. None of this is to say that *Ko Aotearoa Tēnei* is a complete write-off. The discussion it provides of the many issues it covers are important, nuanced, and well-researched, and can provide significant traction in any number of areas. For example, Māori criticisms of the recent TPPA negotiations drew heavily on research and findings from *Ko Aotearoa Tēnei*. Examples such as this show that *taonga* will not easily be put to rest.

Furthermore, the Wai 2358 claim on water has provided a significant breakthrough for *taonga*. Lawyer Jacinta Ruru remarks of the Waitangi Tribunal's findings on the Wai 2358 water claims that they are "the most legally significant Waitangi Tribunal report to date, ever."[48] The Wai 2358 Tribunal found that the *taonga* know in *te ao Pākehā* as "water" is indeed in full ownership of its *kaitiaki* and has been since 1840. Perhaps more significantly, the Tribunal found that the way in which Māori "own" taonga is not to be determined from the West: "That right of property was not constrained by what could be legally owned in England. Rather it depended on what Māori possessed at the time in custom and in fact."[49] With this statement we are taken right to the heart of the Western conceit, that is, property ownership in Aotearoa New Zealand, and a glimpse out of the "one world" – one nature, many cultures – we are currently all forced into, and which *Ko Aotearoa Tēnei* reinstituted. With the Tribunal's response to the water claims, the "silent antagonism" that has hitherto existed between worlds is again challenged, and the "one-world world" looks to give way again to a multiplicity of worlds.[50] Negotiation, diplomacy and discussion appear more important than ever.

NOTES

1. Following some tense encounters, the Waitangi Tribunal was eventually to side with the claimants but again, as with the referendum, the National Government ignored the Tribunal's findings and continued with their privatization programme. The NZMC then initiated judicial review litigation through the High Court, and when this was unsuccessful, they took their case to the Court of Appeal who supported the High Court in finding that there was nothing about the partial-privatization legislation that would contravene Māori rights or impede the Crown's ability to compensate for Treaty breeches.

2. Waitangi Tribunal, *The Stage 1 Report on the National Freshwater and Geothermal Resources Claim* (Wellington: Legislation Direct, 2012), 5.

3. David Williams, "*Ko Aotearoa Tenei:* Crown Law and Policy Affecting Maori Culture and Identity," *International Journal of Cultural Property* 20, no. 3 (2013): 324–25.

4. *Iwi* are large tribal groups that are composed of many subtribes and family groups.

5. The Crown's official response is being led by Te Puni Kokiri, who had placed it in their Statement of Intent in 2012. However, the following year, the Statement of Intent no longer contained any reference to the construction of such a response.

6. Anne Beston, "'Grandfather' of Treaty Claims Nears Final Shape," *New Zealand Herald*, June 16, 2006, http://www.nzherald.co.nz/nz/news/article.cfm?c_id=1&objectid=10386941.

7. Jessica Lai, *Indigenous Cultural Heritage and Intellectual Property Rights: Learning from The New Zealand Experience?* (London: Springer, 2014); Christopher Heath and Anselm Kamperman Sanders, *New Frontiers of Intellectual Property Law: IP and Cultural Heritage, Geographical Indicators, Enforcement, Overprotection* (Oxford: Hart Publishing, 2005).

8. Jonathon Milne, "Maori Want Cut of Research Profits," *New Zealand Herald*, Febraury 27, 2005, http://www.nzherald.co.nz/nz/news/article.cfm?c_id=1&objectid=10112819.

9. Waitangi Tribunal, *Ko Aotearoa Tēnei* (Wellington: Legislation Direct, 2011), 80.

10. Gisborne District Councillor Manu Caddie exclaimed shortly after the reports release that: "This report was expected to provide clarity on property rights for Maori but instead presents a series of schizophrenic findings and recommendations." His frustration was targeted particularly at the report's approach to Maori relationships with flora and fauna, which, he argues, "effectively denigrated Mäori to associated people who have 'important knowledge' with entitlement to a 'reasonable degree of protection' over flora, and fauna" (Caddie). Academic Aroha Mead has said she believes Maori will be "very disappointed in the findings that they don't have proprietary interests in the taonga species, indigenous flora and fauna," and lawyer Moana Jackson, who was heavily involved with the claim, including helping with the drafting of the claim, notes that the Tribunal took a "constrained approach" and considers the recommendations as "better than nothing."

11. Tribunal, *Ko Aotearoa Tēnei*, 78–80, 192–93.

12. Christopher Finlayson, "Government Receives Waitangi Tribunal Report," *Beehive.govt.nz*, July 2, 2011, https://www.beehive.govt.nz/release/government-receives-waitangi-tribunal-report.

13. Kate Chapman and Belinda McCammon, "Govt to Take Time Considering Wai 262 Report," *Stuff.co.nz*, September 4, 2011, http://www.stuff.co.nz/national/politics/5228332/Govt-to-take-time-considering-Wai-262-report.

14. Williams, "*Ko Aotearoa Tenei*."

15. Tribunal, *The Stage 1 Report on the National Freshwater and Geothermal Resources Claim*, 37.

16. Iwi Chairs, "Fresh Water," http://iwichairs.maori.nz/our-kaupapa/fresh-water/.

17. Nick Smith and David Carter, "Fresh Start for Fresh Water," *Beehive.govt.nz*, May 9, 2011, https://www.beehive.govt.nz/feature/fresh-start-fresh-water.

18. Crown Law, "Crown Closing Submissions for Stage One," Wai 2358 Hearings before the Waitangi Tribunal, July 20, 2012, pp. 6, 12–15.

19. Jade Tangiahua Baker, "Te Pahitauā: Border Negotiators," *International Journal of Cultural Property* 15 (2008): 141–57.

20. Mario Blaser, "Political Ontology," *Cultural Studies* 23, no. 5 (2009): 877.

21. Treaty of Waitangi, 1840. Translated by Sir Hugh Kawharau. Accessed February 23, 2015 at http://www.justice.govt.nz/tribunals/waitangi-tribunal/treaty-of-waitangi/the-kawharu-translation.

22. Paul Tapsell, "The Flight of Pareraututu: An Investigation of Taonga from a Tribal Perspective," *The Journal of the Polynesian Society* 6, no. 4 (1997): 326.

23. Ibid., 327.

24. Ibid.

25. Ibid., 328.

26. Ibid., 329.

27. Ibid., 334.

28. Also see A. Salmond, "Nga Hurahi o te ao Maori: Pathways in a Maori World," in *Te Maori: Maori Art from New Zealand Collections* (Abrams: New York, 1984), 109–37.

29. Tapsell, "Pareraututu," 337.

30. Ibid., 338.

31. Marcel Mauss, *The Gift: The Form and Reason for Exchange in Primitive Societies*, trans. W. D. Halls (London: Routledge, 1925), 13.

32. Ibid., 15, emphasis added.

33. Amiria Henare, "Taonga Maori: Encompassing Rights and Property in New Zealand," in *Thinking Through Things: Theorising Artefacts Ethnographically* (London: Routledge, 2007), 48.

34. Ibid., 49.

35. Wai 262 Second Amended Statement of Claim on behalf of Ngati Kuri, Te Rarawa and Ngati Wai, October 20, 2001.

36. Henare, "Taonga Maori," 51.

37. Waitangi Tribunal, "The New Approach" (Wellington: Waitangi Tribunal, 2005), http://www.justice.govt.nz/tribunals/waitangi-tribunal/the-claims-process/the-new-approach.

38. David William is critical of the Tribunal for not exploring these aspects of the claim. He argues that in doing so the Tribunal went straight for the "middle ground" and thus failed to give the report the depth it needed.

39. Tribunal, *Ko Aotearoa Tēnei, Tuarua*, 6.

40. Linda Tuhiwai Smith, *Decolonising Methodologies* (London: Zed Books, 2006), 25.

41. Arun Agrawal, "Dismantling the Divide between Indigenous and Scientific Knowledge," *Development and Change* 26, no. 3 (1995): 431.

42. John Locke, *Two Treatises of Government*, ed. Peter Laslett (Cambridge: Cambridge University Press, 2005).

43. Angelini Massimo, "Down to the Roots of the Word 'Culture'," *PAN: Philosophy, Activism, Nature* 90 (2012): 90–94.

44. Bruno Latour, *We Have Never Been Modern* (Cambridge: Harvard University Press, 1993).

45. Waitangi Tribunal, *Ko Aotearoa Tenei*, 37.

46. Latour, *We Have Never Been Modern*, 97.

47. Michael Goldsmith, "Who Owns Native Nature? Discourses of Rights to Land, Culture, and Knowledge in New Zealand," *International Journal of Cultural Property* 16, no. 3 (2009): 325–39.

48. Jacinta Ruru, "Māori Right in Water: The Waitangi Tribunal's Interim Report," *Māori Law Review*, September 2012, http://maorilawreview.co.nz/2012/09/maori-rights-in-water-the-waitangi-tribunals-interim-report/.

49. Tribunal, *The Stage 1 Report on the National Freshwater and Geothermal Resources Claim*, 140.

50. John Law, "What's Wrong with a One-World World?" *Distinktion: Scandinavian Journal of Social Theory* 16, no. 1 (2015): 135.

BIBLIOGRAPHY

Agrawal, Arun. "Dismantling the Divide between Indigenous and Scientific Knowledge." *Development and Change* 26:3 (1995): 413–39.

Baker, Jade Tangiahua. "Te Pahitauā: Border Negotiators." *International Journal of Cultural Property* 15 (2008): 141–57.

Beston, Anne. "'Grandfather' of Treaty Claims Nears Final Shape." *New Zealand Herald*, June 16, 2006, http://www.nzherald.co.nz/nz/news/article.cfm?c_id=1&objectid=10386941.

Blaser, Mario. "Political Ontology." *Cultural Studies* 23:5 (2009): 877.

Chapman, Kate and Belinda McCammon. "Govt to Take Time Considering Wai 262 Report." *Stuff.co.nz*, September 4, 2011, http://www.stuff.co.nz/national/politics/5228332/Govt-to-take-time-considering-Wai-262-report.

Crown Law. "Crown Closing Submissions for Stage One." Wai 2358 Hearings before the Waitangi Tribunal, July 20, 2012, pp. 6, 12–15.

Finlayson, Christopher. "Government Receives Waitangi Tribunal Report." *Beehive.govt.nz*, July 2, 2011, https://www.beehive.govt.nz/release/government-receives-waitangi-tribunal-report.

Goldsmith, Michael. "Who Owns Native Nature? Discourses of Rights to Land, Culture, and Knowledge in New Zealand." *International Journal of Cultural Property* 16:3 (2009): 325–39.

Heath, Christopher and Anselm Kamperman Sanders. *New Frontiers of Intellectual Property Law: IP and Cultural Heritage, Geographical Indicators, Enforcement, Overprotection*. Oxford: Hart Publishing, 2005.

Henare, Amiria. "Taonga Maori: Encompassing Rights and Property in New Zealand." *Thinking Through Things: Theorising Artefacts Ethnographically*. London: Routledge, 2007.

Iwi Chairs. "Fresh Water." http://iwichairs.maori.nz/our-kaupapa/fresh-water/.

Lai, Jessica. *Indigenous Cultural Heritage and Intellectual Property Rights: Learning from The New Zealand Experience?* London: Springer, 2014.

Latour, Bruno. *We Have Never Been Modern*. Cambridge: Harvard University Press, 1993.

Law, John. "What's Wrong with a One-World World?" *Distinktion: Scandinavian Journal of Social Theory* 16:1 (2015): 126–39.

Locke, John. *Two Treatises of Government*. Ed. Peter Laslett. Cambridge: Cambridge University Press, 2005.

Massimo, Angelini. "Down to the Roots of the Word 'Culture'." *PAN: Philosophy, Activism, Nature* 90 (2012): 90–94.

Mauss, Marcel. *The Gift: The Form and Reason for Exchange in Primitive Societies*. Trans. W. D. Halls. London: Routledge, 1925.

Milne, Jonathon. "Maori Want Cut of Research Profits." *New Zealand Herald*, Febraury 27, 2005, http://www.nzherald.co.nz/nz/news/article.cfm?c_id=1&objectid=10112819.

Ruru, Jacinta. "Māori Right in Water: The Waitangi Tribunal's Interim Report." *Māori Law Review*, September 2012, http://maorilawreview.co.nz/2012/09/maori-rights-in-water-the-waitangi-tribunals-interim-report/.

Salmond, A. "Nga Hurahi o te ao Maori: Pathways in a Maori World." *Te Maori: Maori Art from New Zealand Collections*. Abrams: New York, 1984, 109–37.

Smith, Linda Tuhiwai. *Decolonising Methodologies*. London: Zed Books, 2006.

Smith, Nick and David Carter. "Fresh Start for Fresh Water." *Beehive.govt.nz*, May 9, 2011, https://www.beehive.govt.nz/feature/fresh-start-fresh-water.

Tapsell, Paul. "The Flight of Pareraututu: An Investigation of Taonga from a Tribal Perspective." *The Journal of the Polynesian Society* 6:4 (1997): 323–74.

Treaty of Waitangi. 1840. Translated by Sir Hugh Kawharau. Accessed February 23, 2015 at http://www.justice.govt.nz/tribunals/waitangi-tribunal/treaty-of-waitangi/the-kawharu-translation.

Waitangi Tribunal. "The New Approach." Wellington: Waitangi Tribunal, 2005, http://www.justice.govt.nz/tribunals/waitangi-tribunal/the-claims-process/the-new-approach.

———. *Ko Aotearoa Tēnei*. Wellington: Legislation Direct, 2011.

———. *The Stage 1 Report on the National Freshwater and Geothermal Resources Claim*. Wellington: Legislation Direct, 2012.

Williams, David. "*Ko Aotearoa Tenei:* Crown Law and Policy Affecting Maori Culture and Identity." *International Journal of Cultural Property* 20:3 (2013): 324–25.

Part II

CULTIVATION AND CULTURE

Chapter 5

The Plough as Settler Colonial Cultural Icon

Voices from the Other *Side of the Blade*

Victoria Grieves

Australian, Aotearoa New Zealander and North American settler colonial societies are marked by an ongoing extraordinary enthusiasm for the plough that began during the nineteenth and early twentieth centuries. This enthusiasm is allied to a zeal for the western patriarchal ideal of "men's mastery over nature." This has seen the original colonial conquest underlined and made more thoroughgoing by the fact that not one part of the natural world is now unaffected by the activity of humans. We are faced with the increasing spectre of the complete destruction of the natural world, in this the age of the Anthropocene.[1] The plough has become a central, iconic symbol of settler colonial values and ethics that were developed *against* everyday constructions of Indigenous peoples primitivity, ignorance and backwardness. In fact, the plough became a *weapon in the war* against nature and Indigenous custodians.[2]

The focus of this chapter is the iconic idea of the plough in the settler colonial cultural imaginary, especially in the period of its greatest take-up as a tool of colonization in nineteenth and early twentieth century Australia, Aotearoa New Zealand, Canada and the United States. The focus is also on the voices of those who have themselves been alienated by the action of the plough; Indigenous peoples and other *outlaws*, whose struggle has sometimes focused on reinscribing and unsettling the idea of the plough as a colonial imposition within the historical development of settler colonial societies. Concomitantly, other newcomers to Indigenous lands were seduced by the poetry of the plough, the idea of good honest toil bringing just rewards, and a failsafe way to become *settled*. The extent of the devastation wrought by the plough including for example, deforestation and global dust storms, are allied to the thinking of Michel Serres. He grounds the plough as a seemingly irrepressible phenomenon of man's war with nature that has led the world into what we now recognize as the Anthropocene.[3] Considerations for decolonization of Indigenous people and lands must engage with the iconic status of the plough in western society and find ways to reinscribe the land use patterns of Indigenous peoples as we move further into the Anthropocene.

I come to this work as one of the people who know themselves to be Warraimaay, whose country is located in the midnorth coast of New South Wales (NSW), Australia, and whose custodianship extends for the last fifteen thousand years or more. Thus I am a Warraimaay historian in the academy, working in Indigenous knowledge developments from a baseline of Aboriginal philosophy.[4] As a child I was with my mother, also Warraimaay, who worked on pastoral stations in the country of the Gamiraay and Wiradhuraay peoples in western and south-western NSW respectively, and I became aware of the human impact on that country through wholesale clearing of vegetation and ploughing of the soil.[5] As an Aboriginal person I am dedicated to untangling the myriad settler colonial impacts on people and country that have led us to the predicaments that we are in, in this age of the Anthropocene. While I recognize important cultural differences among Indigenous peoples globally, I also understand that our cultures hold common values and ethics in relationship to the natural world and the Earth as a living organism; we understand the mystery of phenomena and connections in the natural world and through ceremony, move closer to this mystery. We also hold in common histories of colonial takeovers of our lands in the name of (western) progress and we have our own critical views of this phenomenon. In this connection, critical Indigenous theory has synergies with the new materialist framework developed from within cultural studies and has led me to engage more thoroughly with contemporary western epistemologies.

For those of us engaged in Southern Theory and the importance of the Southern Cross constellation to the knowledges of the Indigenous people of the South of the globe, it is instructive indeed to know that the Plough has been inscribed in the constellations by the civilizations of the north.[6] It is a prominent formation of seven stars in the constellation Ursa Major, the Great Bear, containing the Pointers that indicate the direction to the Pole Star. Importantly, this constellation is also called the Big Dipper in the native cosmologies of North America, as it is the shape of the implement used to serve water and food. The provisioning of the people, the capacity to serve, also includes responsibility for the custodianship and continuing relationships to the natural world that in turn ensure the continuing bounty of the natural world. The symbol of the Plough, iconic to the dominant western civilizations of the north of the globe, has been reinscribed over the Big Dipper. It has no real foothold in the cultures of the Indigenous peoples as they are now known, those people whose cultures have been born of, and built on, connectedness and relatedness to the natural world.[7] In this way they are peoples who retain the valuable prescriptions for living within nature. Thus the plough can be analysed as an imperfect and improper imposition on the Indigenous peoples of the global South, resonating with Serres' idea of the plough as a weapon against nature.

The plough began as a wooden digging stick drawn through the earth to create a neat channel for the growing of seed, a delicate touch in comparison to the nature of ploughing developed in the last two hundred years. The nineteenth century saw a flurry of innovation with the opening up of new lands for cultivation in the colonies and high demand for the implements of ploughing. In response, the plough developed during the nineteenth century so as to be almost unrecognizable from the Rotherham swing plough of the seventeenth century or the Scots plough of the eighteenth century. Ploughs were pulled by people, horses and oxen or even by steam units in some contexts by the end of the nineteenth century.[8] The metal, multifurrowed machine developed out of settler colonial conquests of Indigenous lands and is now ubiquitous in the world of agriculture. It has been overwhelmingly championed within settler colonial societies for taming the land and bringing civilization.

Ploughing in England, Scotland and other European societies that have cold and temperate climates, goes no deeper than twenty centimetres, a depth widely understood and accepted to be beneficial for crop yield. The argument is that the action of ploughing in furrows, turning the topsoil, has an aerating and manuring effect that controls weeds and common pests and facilitates the seeding of new crops. There are also cold and temperate environments suited to ploughing of the soil for crops in North America, including Canada, Aotearoa New Zealand and to a lesser extent in Australia. However, the practice of ploughing the soil was introduced into marginal agricultural lands in Australia, north and South Africa and the west and northwest of the United States, where it is referred to as dry homesteading. These drylands suffer water stress, they are arid and semi-arid and dry subhumid lands[9] and the results of ploughing these lands continue to be devastating.[10]

THE PLOUGH AS CULTURAL ICON: REPRESENTING FAITH AND VIRTUE

The promotion of drylands farming was ubiquitous to the settler colonial societies of North America and Australia in the time of their establishment.[11] Farming was promoted as every migrant white man's right of opportunity and the concept of the farmer carried great social capital as an important foundational profession in the colonies. An example of the promotion of farming and ploughing as honest work that produces good men and families and therefore good societies is to be found in the farming manuals of Herbert Campbell, published in the United States and available in Australia, who referred to the business of farming as "the calling … of millions of the best people of the earth."[12] Campbell's romantic and idealized vision is of the farmer as the basis of a good and prosperous society:

> And the small farmer is the one who makes his farm his home. He seeks comfort for himself and his children. He does not build a shed to shelter him during the crop season with his family miles away. He becomes a permanent fixture in his country. He builds good houses and barns, he gets the best cattle and horses and hogs, he has a garden of flowers and he plants trees. He wants the schoolhouse to be located not far away and he willingly taxes himself in support of the school. He contributes to the erection of a church in the village and he is careful that the rural route and the co-operative telephone do not pass him by. The ideal farmer makes the ideal farm and in return there is compensation enough.[13]

The problem for the willing famers who took up the new lands opened up in the northwest of the United States including Nebraska and the great Plains, in the western regions of NSW, the northern regions of South Australia and the eastern lands of Western Australia for example, was that the land was marginal drylands and the topsoils at risk of total destruction through ploughing.

The plough had become so iconic to settler colonial societies during the nineteenth century that there was no alternative but to plough. Strange as it may seem, this was especially the case when it came to dryland farming as it was embarked upon in the late nineteenth and early twentieth centuries. In fact the misinformed but widely held adage that "Rain follows the plough" came out of the settler colonial need to believe in the power of ploughing to ensure the success of crops.[14] Charles Dana Wilbur a land speculator in the trans-Mississippi northwest of the United States was influenced in his belief that the plough itself produced rain, by earlier agronomists. He wrote:

> Suppose (an army of frontier farmers) 50 miles, in width, from Manitoba to Texas, could acting in concert, turn over the prairie sod, and after deep plowing and receiving the rain and moisture, present a new surface of green growing crops instead of dry, hard baked earth covered with sparse buffalo grass. No one can question or doubt the inevitable effect of this cooling condensing surface upon the moisture in the atmosphere as it moves over by the Western winds. A reduction of temperature must at once occur, accompanied by the usual phenomena of showers. The chief agency in this transformation is agriculture. To be more concise, *Rain follows the plow*.[15]

The driver that produced devastation of lands *opened up* for farming by the action of the cutting and turning of the soil was primarily a deep cultural belief and faith in the progress brought about by English-style farming that persisted, even during failure after failure in farming marginalized lands.

THE WRONG SIDE OF THE BLADE: THE PLOUGH AS WEAPON AGAINST PEOPLE

The plough is also an implement of human bondage and servitude though it is seldom cast in this mould. Humans have often pulled it as slave labour in old worlds and new, but particularly in the early years of the spread of ploughing to the colonized new worlds. In Australia, the first settlements saw convicts used as beasts of burden until the large-scale adoption of bullocks and later horses. This is reflected in the convict ballad *Van Diemens Land*:

The first day we landed upon this fatal shore
The planters they came round us, full twenty score or more
They rank'd us up like horses, and sold us out of hand,
And yok'd us up to ploughs, my boys, to plough Van Diemen's Land.[16]

The outlaw Ned Kelly was aggrieved that his own father was one of the men used in this way. In his captivating essay on the bushranger Ned Kelly's *The Jerilderie Letter* of 1879, the poet Michael Farrell explains that the plough is closely associated with poetry in western literature where such connection is "obvious and inevitable."[17] Kelly was not literate and narrated some 8,000 words to his friend and accomplice Joe Byrne who wrote what emerges as a poetic rationale for all of Kelly's actions on the wrong side of the law (or the wrong side of the blade). Farrell rereads what has been understood purely as a personal and political text of an Australian bushranger, for Kelly's deep engagement with the idea of the plough, his resentment of the agricultural milieu from which it sprang and that Kelly sought to unsettle. Kelly dismantled the assemblage of agriculture and ploughing in his life works – his bushranging and his written work.

Kelly came from a family for which the small-scale farming endeavour had not been successful, they endured one calamity of nature and issues with local authorities after another. Farrell establishes the settler poetics of the plough as aligned to the deep yearning of the settler colonial to measure up to the culture of the old country, such as is evident in the work of Dame Mary Gilmore (1865–1962). Acutely aware of perceived shortcomings of the colonies' culture in comparison to that of the metropole, she lamented, "Why haven't we our great fiction … of the hand that guided the plough?"[18] Thus the plough is iconic to colonization, representative of burden and toil as well as class privilege, and it is this that Kelly railed against.

In this connection, it was Kelly who reinscribed the nature of the plough by destroying the plough, as it was understood; he refashioned the metal ploughshare moulds into armour to protect him against the bullets fired by police, representatives of the colonial state.[19] Thus he attempts to physically and symbolically turn the destructive force of the plough against the elites, whose hand would guide the plough, if only to cleave the saints from the sinners. Farrell encourages us to see the hand of Kelly as that which in fact destroyed the plough, and in so doing developed *The Jerilderie Letter*, a great Australian piece of literature, though not of the literary class that Gilmore was referencing. Kelly is one of the *outlaws* who fell on *the other side* of the ploughshare. In like fashion to the other bushrangers or outlaws of this time, in both Australia and in the United States, he eschewed the tilling of land for a living. The establishment rejects him and he also rejects the society that has grown from the plough.

This is a similar tale to other outlaws who began their working lives in agriculture, such as the mixed-race Aboriginal, Jimmy Governor. Though Governor is not known as a ploughman, he was fencing for the squatters who were continuing to plough the land.[20] The story of Governor's actions as an outlaw has become almost mythological in the Australian imaginary, representing the deep psychological fear of the Aboriginal potential for striking back. Governor went on a murdering rampage seeking vengeance from all the settler colonial farmers (ploughmen) who had wronged him. His actions parallel those of Kelly and similarly, his "anti-colonial moment" can be read as a wholesale rejection of the lifeways that accompanied the ploughing of the soil.[21]

What Kelly the bushranger has to say about the plough and its impact on his life and that of his family is potentially representative of all outlaws of his time. Those people whose lives had been shaped by the workings of the plough and who, like Kelly, found themselves on the wrong side of the blade. As Farrell observes, while Kelly was not to be the master of the plough "his detour was no more deadly than that of agriculture if we accept philosopher Michel Serres' argument that the slicing of the earth is a killing."[22]

THE PLOUGH AS WEAPON AGAINST NATURE

Michel Serres' critical analysis brings a new dimension to the destructive force of the plough through understanding its impact on nature. He maintains that there is a reality on earth of which we are not aware, that we do not inhabit a sphere that is of our culture alone (as humans can easily assume we do) and it follows that we are not aware of the impact of our actions on the natural world.[23] Serres comes to an understanding that at this stage of human history nature has become a protagonist in human culture inasmuch as human history has begun to act on nature. In this way then, the separation of the natural and the cultural world of humans, of the sciences and the humanities of "things and signs" becomes inconceivable.[24] He recognizes the need for a new kind of knowledge production that breaches this separation between humans and the natural world, one that can potentially be filled by Indigenous knowledge production and/or the environmental humanities. Important too is that Serres' thinking foregrounds the opposition *within* the Anthropocene; that the activities of humans on the earth including the rise of agriculture and the industrial revolution, have a significant global impact on the earth's ecosystems also means that for the first time, "in this little sliver of time," the forces of nature are impacting on human culture. In his work *The Natural Contract*, he states, "Global history enters culture; global culture enters history: this is something utterly new in philosophy."[25]

Serres continual juxtaposition of the hard and the soft, of science and culture, the physical as opposed to the conceptual, of things and signs, has him understand the action of the plough, the cutting through the earth, as a killing.[26] In his book *The Parasite* he describes an epiphany – while considering the impact of the Yellow River on the countryside he suddenly understands the impact of agriculture on the natural world – "I think it is radical – rigorously, literally radical" – as tearing out by the roots to lay bare the ground for planting. He describes the development of a field, the denuding of the earth, of the uprooting of existing plants as if in a battlefield, the blade of the plough a sacrificial blade that shapes space and the earth, that "does not stop in its murderous way," redefining what will live and not live.[27] The plough is more than an agent of colonization, death and destruction; it is an arbiter of good and evil. Serres describes how the ploughshare can only slice indiscriminately, without thinking, and in so doing defines the sacred and the profane, the temple and the area of evil. While the plough has cut a line of human *righteousness* through the world, Ned Kelly and his ilk have come up on the wrong side of the furrow, along with the vast majority of the species of the natural world and Indigenous peoples.

"MY WEAPON WAS THE PLOUGH!"

The Taranaki Māori of the Indigenous people of Aotearoa New Zealand have employed the plough as a means of passive protest, to unsettle the "settlement" of usurpers the *pakeha*. The Taranaki Māori, the people of the sacred ancestral mountain Taranaki, appropriated the plough and used it against the plough's masters as a way of demonstrating the impact of having land ploughed against one's wishes for the takeover of lands. An anonymous Māori ploughman of the time is quoted as saying, "My weapon was the plough, but yours was firearms."[28]

The decision to plough the land appropriated by the white man as a form of passive resistance was made from deep within *Māori* cultural values by chiefs such as Te Pokiha Omaharu and Te Whiti O Rongomai who was a spokesperson for the people.[29] Another leader was Wiremu Kingi Te Rangitake who addressed the problem of when *pakeha* ploughed land, they then desired to claim more Māori land beyond what had already been acceded.[30] He had long refused to have government surveyors come onto Māori land and had not utilized any violence in this request to stay away. In response to this refusal, the government troops fired on Te Ratangi's *pa* at Te Kohia in 1860. Te Ratangi and surrounding Māori rallied and were victorious at the Battle of Waireka thirteen days later. The very next day the government bombarded the village of Warea, that contained women and children, from battleships staged off the coast. This was closely followed by an assault by ground troops that strangely avoided the *pa*. The Māori casualties were devastating. As a result of the overwhelming firepower and radically unethical tactics of the enemy, the leader Te Whiti and other chiefs decided to make a non-violent stand. They redefined and inverted the settler colonial instrument of destruction, the plough, to remake it as a means of passive resistance in the face of a threatened "death blow" from the new settlers of Parihaka.[31]

The story of events at Parihaka has become an important, celebrated example of resistance for those whose hands only guided the plough in protest. *Māori* oratorical skill including the singing of stories of events from their history is responsible for the retention of this knowledge and important too is the diary of Erenora a Parihaka woman. When in her eighties she narrated the story of these turbulent times and the manuscript, written in Māori was held in an archive, overlooked and all but forgotten. Another account of the times, *I Te Raa o Maehe* had been collected from the man who composed it: "Tonga Awihikau, [is] a returned ploughman who died in 1957 aged 104. He is remembered for leading the land struggle in the 20th century. People flocked to hear his oratorical skills and to see this charismatic tamaiti rangatira noo Taranaki (child leader of Taranaki)."[32] Thus it was, that people with their backs to the wall, Indigenous people faced with all the might of the British Empire, appropriated and utilized the iconic settler colonial plough for the purposes of resistance. Their understanding of the plough as the central driver of the colonization of their lands is evident; the plough itself brought death and destruction and forced them to adapt to the reality of a new set of imperatives in their lives.

THE WORST HARD LESSON TO LEARN

The creation of the great dustbowl in the United States has been referred to as the "worst hard time."[33] In spite of many dustbowls and dust storms over approximately two centuries, the lesson about the downside of the use of the plough has been the hardest for the settler colonial to learn. Early warnings of the extent of devastation likely to occur, such as those of George March an early American environmentalist in 1865, were ignored. He warned of the dangers of large-scale deforestation for the development of plough lands. His concerns included the impact of deep ploughing in particular tends to destroy the whole plant life, roots and all; there is increased evaporation from the soil and less water holding quality. With no means of holding the soil together the expansion of the shifting sand deserts of the world becomes inevitable.[34] His alarm and concern for the future of the environment seems to have had little impact on the takeover of virgin lands in settler colonial contexts in the subsequent decades.

The farming of marginal lands, ploughing and overgrazing, exacerbates the process of desertification. This knowledge has been gained most graphically through the bitter experience of the Great Dust Bowl in the west and northwest of the United States during the 1930s. Dust storms, dust clouds, dust pneumonia destroyed stock and killed people in their youth, dust that had to be scooped up from within the houses each day and dust covered the sun producing an eerie darkness. Chickens went to roost mid-afternoon, people tethered themselves by ropes so as not to get lost when moving between farm buildings. Dust clouds rose up hundreds of feet in the sky and rolled like storm clouds. "A storm in May 1934 carried the windblown shards of the Great Plains over much of the nation. In Chicago, twelve million tons of dust fell. New York, Washington – even ships at sea, three hundred miles off the Atlantic coast – were blanketed in brown."[35] The destructive outcome of ploughing and "breaking the sod" of the Great Plains has not been able to be ignored. However, while this phenomenon has been examined to such an extent it has entered popular culture, the subsequent never-ending dust storms around the world are not often reported. In Australia, for example, successive dust storms began with the "Rain of Blood" over Melbourne in 1896 and have continued with an increase of large rolling dust storms that travelled thousands of kilometres between 1920 and 1950.[36]

The Great Plains had long been the preserve of the Comanche, custodians of a unique ecology; high grass plains able to withstand the searing heat of summer and the snow and ice of winter. A treeless plain of tall, deep-rooted grasses that kept the soil in place, trapping moisture even in periods of drought and high winds. The Comanche are people worthy of the Deleuzian imprimatur of the nomads' *war machine*; they moved in circumscribed patterns in country that they absolutely knew. They were built for battle and proved undefeatable for many decades. It was only after the end of the Civil War and the development of more efficient armoury and war-crazed veterans, with no thought for life or limb to pursue them, that they were defeated in 1875.[37]

While the war that Deleuze and Guattari refer to is more of a symbolic battle, that of becoming over being, of the nomadic over the sedentary, it can be argued that the Comanche were engaged in this battle as well as in the physical battle of war. As a people they were inextricably tied to the Great Plains so as to be a part of the landscape, they were attuned to the life of the environment, they were continually "becoming" with this landscape. "The question is therefore less the realization of war than the appropriation of the war machine. It is at the same time that the State apparatus *appropriates* the war machine, subordinates it to its 'political' *aims*, and gives it war as its direct *object*."[38] Thus the Comanche eventually succumbed to the settler colonial state, and were defeated, as were the majority of Indigenous people of the world.

It is important that the bloody and brutal colonization of the Indigenous peoples of settler colonial Australia, Aotearoa New Zealand and the Americas was largely brought about by the imperative of the settler colonial to plough. Ploughing was an important weapon in the war against nature, to erase and transcend the natural realities of rural places.[39] As Indigenous custodianship of country was erased, the wholesale destruction of the biodiverse landscapes occurred. George Main describes how it was the carefully managed and developed natural landscapes of Wiradhuraay custodianship that were destroyed on the southern tablelands of NSW in a matter of a decade.[40]

THE CALAMITY OF DUST CLOUDS

It was the rapid mechanization of farm implements, especially the introduction of small gasoline tractors, sod breaking ploughs and the combine harvester that destroyed the ecology of the Great Plains. In Nebraska, for example, the number of gasoline tractors increased by almost 90 percent between 1919 and 1920.[41] These innovations led to attempts to convert this arid grassland, much of which received no more than 10 inches (250 mm) of precipitation per year, to cultivated cropland. The extensive deep ploughing of the Great Plains reached a zenith in the decade preceding the Great Dust Bowl and saw the destruction of the natural deep-rooted grasses. It was the usual drought and high winds of this dryland country that then resulted in the topsoil being blown away over some decades.

In Australia, as in the United States, the threat of another dust bowl has never ceased. It is the dust of the countryside that sometimes makes a huge impact in the cities and towns. Just one of the examples of this phenomenon occurred in September 2009 when a dust cloud moved across NSW, Queensland and the Australian Capital Territory measuring 3,450 kilometres, carrying some 16 million tonnes of dust from Central Australia. At the peak of the storm the Australian continent was losing 75,000 tonnes of dust per hour off the NSW coast north of Sydney. Visible from space, this event was said to be the worst in the state of NSW for nearly seventy years, from the days of the original dust bowl phenomenon.

In the drylands of Australia as in the Great Plains, settler colonials constructed the natural world and the people in it, Aboriginal people and their land management practices, as a barrier to production. George Main has argued that the discipline of agricultural science was based on the belief that "humans can defy natural limits and boost primary production through the simplification and strident transformation of local ecologies." Thus the natural characteristics of the physical environment of rural Australia were cast as problematic barriers to production – "even the location of the continent an obstacle to successful agriculture."[42]

Main describes the settler colonial engagement with the land as a kind of warfare that utilizes more and more efficient weapons. Single tractors with the pulling power of 500 horses are advertised for warfare "like the legendary Knights of old – commanding respect – waiting to be unleashed – can do battle in the harshest conditions." Moreover, Main argues, the idea of *battling the land*, "reveals and reinforces heroic narratives of rural colonisation" whereby conflict between settler colonials and the natural world lies at the heart of understanding the true story of rural Australia.[43] A rural Australia where the cry of *God speed the plough!* was met with cheers at the end of many a politician's speech.[44]

In 2005, the *Millennium Ecosystem Report on Desertification* established that drylands, which range from "dry sub-humid" to "hyper-arid" regions, make up more than 40 percent of the world's land surface and are home to two billion people. The largest area stretches from Saharan Africa across the Middle East and Central Asia into parts of China. Most of Australia is also classified as drylands, along with much of the western United States, parts of southern Africa, and smaller areas of desert in South America. The report states that up to 20 percent of those areas had already suffered some loss of plant life or economic use as a consequence of desertification. Desertification has also been linked to health problems, poverty and a drop in farm production, with infant mortality in drylands double the rate elsewhere in developing nations. Moreover, the spread of deserts causes changes to the environment on a global scale; dust storms in the Gobi and Sahara deserts cause respiratory problems as far afield as in North America as well as damage to coral reefs in the Caribbean. Scientists estimate that a billion tons of dust from the Sahara are lifted into the atmosphere each year.[45]

CONSIDERATIONS FOR DECOLONIZATION

In the film *Interstellar* (2014), human beings' time on earth is coming to an end as it no longer supports and sustains them. A team of explorers undertakes what is billed as the most important mission in human history, travelling beyond this galaxy to discover whether mankind has a future among the stars – on a planet that is capable of supporting human life. The reason why they have to leave earth in this way is a scenario that is by now deeply embedded in the cultural imaginary: the topsoil of the earth is blown away, immense dust storms threaten human life and agricultural yields are dwindling to become almost non-existent. Various food crops are attacked by blight, one by one, and there have been extreme food shortages. The film employs a faux documentary device of interviews with people telling of the development of the dust phenomenon on earth. These are actual interviews with survivors of the great dust bowl of the United States in the 1920s and 1930s such as those in documentaries made of the Great Dust Bowl, that have been shown on television in the United States and here. The informants speak of the dust covering everything even the dinner table where the plates and cups have to be set upside down, the necessity of wearing strips of cloth over the nose and mouth in an attempt to stop the dust entering the lungs. A woman says: "You didn't expect this dirt, that was giving you this food, to turn on you."

The film introduces the debate about why there is a necessity for exploring outer space for a new colony, but this is not fully explored. In the world of Interstellar, an anti-space science policy has developed in that there is a push to look down instead of up, to focus on fixing the earth rather than reaching for the stars. The question is posed: Does the world now need farmers more than it needs engineers? The protagonist, a space scientist and pilot, says: "We used to look up at the sky and wonder about our place in the stars, now we only look down and wonder about our place in the dirt."

In a conversation about the necessity for space travel to find another place for people to live a scientist says, "We have always been driven by the unshakeable faith that the earth is ours" the response "well not just ours no, but it is our own." The consensus reached within the film is to take a long shot – to find a home, to establish a new colony. "We are explorers, this is our boat." This raises the question of the true nature of colonization, the finding of new homes for "us," at a certain level, outside of the petty differences we have among us, one organism. However, for Indigenous peoples the Earth is always our home no matter how it is cast and no matter what damage is done, the imperative is to work to retain, reclaim and rebuild. Our mission as Indigenous peoples is to keep every component of the natural world alive.

CONCLUSION

As Serres so aptly puts it in *L'Incandescent*:

> I have never known how nor been able to separate myself from realism, hard, for the idealists, soft, seem to me never to have suffered from the world as such; raised in cotton wool, coddled and protected, the rich, the powerful and their children believe that all the things of the world obey them like their servants.[46]

This quote encapsulates the challenge for the people of the Anthropocene: to humble ourselves, become cognizant of that world that exists under our gaze but at the same time just out of our direct line of vision, and seek to preserve that world.

This quest could well be an article of faith, as much as it is of hard science, as it is for Serres and for myself. Paradoxically, it is an article of faith for western settler colonials to continue to plough the soil, even after the evidence of degradation is clearly all around them. Perhaps people who are born into new landscapes come to see them as normal and natural ways for the Earth to be and cannot imagine a different kind of management with different outcomes. George Main has discovered and pieced together a puzzle of former Wiradhuraay managed landscapes in his *Heartland*, that were essentially wet and moist, teeming with life. The country now is dry, dry, dry and droughts dog the unwitting farmers whose cultural load in fact produced them.[47] In this we find ourselves in a struggle of cultural imaginaries, whereby the recent film *Instellar* unquestioningly repeats the great and good imperative to plough the soil and to continue to find other lands in which to plough the soil, in the face of fast fading opportunities for human life on Earth. Indigenous peoples are content to look down, on the Earth, and find our place on it, within the workings of the natural world.

I believe it is the world of the ritual of Indigenous peoples that connects humans to the natural world and the eternal mystery of creation. That state between wakefulness and sleep is closest to the eternal, as the *Arrente* elders explained to the anthropologist Stanner almost a century ago.[48] That is the world of eternal time that travels along beside us in the present, that is "everywhen," the world of paradigms for human behaviour that are also written in the constellations, the maps that guide the Indigenous people of the earth in the preservation of the natural world. It is this world that has been untouched by the blade and still exists; it is now the time to take stock of what we have retained and can retrieve. George Main describes the activities of farmers in the country he calls his *Heartland*, people who are seeking to regrow the natural grass and woodlands, to preserve the moisture in the soil, establish sanctuaries for species; building and providing careful custodianship patterned on that of the Wiradjuri peoples that has nurtured and provided succor to *country*.[49] This praxis addresses the need to reject, repair and resolve the impacts of ploughing on the natural world and develop new ways of relating to the Earth that respect and incorporate the enduring and now increasingly important wisdom of Indigenous custodians.

NOTES

1. Joseph Stromberg, "What is the Anthropocene and Are We in It," *Smithsonian* Magazine, January 2013, http://www.smithsonianmag.com/science-nature/what-is-the-anthropocene-and-are-we-in-it-164801414/?no-ist; Adam Vaughan, "Human Impact Has Pushed Earth into the Anthropocene, Scientists Say," *The Guardian*, January 7, 2016.
2. George Main, *Heartland: The Regeneration of Rural Place* (Sydney: University of New South Wales Press, 2005), 50–51.
3. Michel Serres, *The Parasite*, trans. Lawrence R. Schehr (Baltimore: John Hopkins Press, 1982).
4. Victoria Grieves, *Aboriginal Spirituality: Aboriginal Philosophy: The Basis of Aboriginal Social and Emotional Wellbeing* (discussion Paper no. 9, Darwin: Cooperative Research Centre for Aboriginal Health, 2009), http://www.crcah.org.au/publications/DP_series.html.
5. Victoria Grieves, "Narranga Barrangang: Self and History, a Contemporary Aboriginal Journey," in *Ngarpartji Ngarpartji: Ego-Histoire from Europe and Indigenous Australia*, ed. Cole, Haag, Castejon and Hughes (Canberra: ANU ePress, 2014).
6. Particularly Wanta Jampitjinpa Patrick, senior lawman of the Walpiri people in the Northern Territory who is instructing me on the importance of the Southern Cross to understanding the impact of the emu and other non-flying birds of the south of the globe and the impact of the bear in the north of the globe. I am also indebted to other Indigenous colleagues in the developing Southern Cross Dialogues: Jalani Neufville Niaah and Sonjah Stanley Niaah of the University of the West Indies; Genner Llanes Ortiz of Lieden University.
7. Grieves, op.cit., 15.
8. Alan Jones, "A Brief History of the Plough," *Society of Ploughmen* (n.d.), http://www.ploughmen.co.uk/about-us/history-of-the-plough.
9. Food and Agricultural Organisation of the United Nations (FAO), "Drylands People and Land Use," *Waters and Cereals in Drylands* at (London: Earthscan, 2008), 1, http://www.fao.org/docrep/012/i0372e/i0372e01.pdf.
10. Main, *Heartland*; Cameron Muir, *The Broken Promise of Agricultural Progress: An Environmental History* (London: Routledge Environmental Humanities Series, 2014), 111.

11. Muir, *Broken Promise*, 99–100.

12. Hardy Campbell, *Campbell's 1907 Soil Culture Manual: A Complete Guide to Scientific Agriculture as Adapted to the Arid Regions* (Lincoln, NE: The Campbell Soil Culture Company Inc., 1907), 23, http://archive.org/details/campbellssoilcul00camprich.

13. Ibid.

14. Timothy Egan, *The Worst Hard Time: The Untold Story of those Who Survived the Great American Dustbowl* (Boston: Houghton Mifflin Harcourt, 2006), 24.

15. Charles Dana Wilber, *The Great Valleys and Prairies of Nebraska and the Northwest* (Omaha, NE: Daily Republican Printing Company, 1881).

16. Quoted in Michael Farrell, "A Poetics of the Plough: Ned Kelly's *Jerilderie Letter*," *Southerly* 72, no. 2 (2012), 78.

17. Farrell, "Poetics," 78.

18. Ibid., 77.

19. Ibid.

20. Laurie Moore and Stephan Williams, *The True Story of Jimmy Governor* (Crows Nest: Allen and Unwin, 2001).

21. Marilyn Wood, "The Breelong Blacks," in *Race Matters: Indigenous Australians and "Our" Society*, ed. G. Cowlishaw and B. Morris (Canberra: Aboriginal Studies Press, 1997); Moore and Williams, *True Story*.

22. Farrell, "Poetics," 77.

23. Michel Serres, *The Five Senses: A Philosophy of Mingled Bodies (I)*, trans. Margaret Sankey and Peter Cowley (London: Continuum, 2008), 102, 103.

24. Steven Connor, "Michel Serres: The Hard and the Soft" (lecture, Centre for Modern Studies, University of York, 2009), http://www.stevenconnor.com/hardsoft/.

25. Michel Serres, *The Natural Contract*, trans. Elizabeth Macarthur and William Paulson (Ann Arbor: University of Michigan Press, 1995), 4.

26. Serres, *Parasite*, 177; Farrell, "Poetics," 77.

27. Serres, *Parasite*, 177.

28. Te Miringa Hohaia, Gregory O'Brien, and Lara Stringman, *Parihaka: The Art of Passive Resistance* (Wellington: Victoria University Press, 2006), 28.

29. Hohaia, O'Brien, and Stringman, *Parihaka*, 12.

30. Witi Ihimaera, *The Parihaka Woman* (London: Vintage, 2011).

31. Ihimaera, *The Parihaka Woman*.

32. Ibid.

33. Egan, *Worst Hard Time*.

34. George Perkins March, *The Earth as Modified by Human Action* (New York: Charles Scribner, 1865), 141, 194, 231–32, 272, 276, 323, 467.

35. Egan, *Worst Hard Time*, 5.

36. Muir, *Broken Promise*, 109–10.

37. Egan, *Worst Hard Time*, 15–19.

38. Gilles Deleuze and Felix Guattari, *A Thousand Plateaus: Capitalism and Schizophrenia* (Minneapolis: University of Minnesota Press, 1987), 420.

39. Main, *Heartland*, 44.

40. Ibid., 16–56, 108.

41. J. C. Olson, J. J. Montag, and R. C. Nagle, *History of Nebraska* (Lincoln: Nebraska University Press, 2015).
42. Main, *Heartland*, 43.
43. Ibid., 51.
44. Ibid., 62.
45. Millennium Ecosystem Assessment, *Ecosystems and Human Well-being: Desertification Synthesis* (Washington, DC: World Resources Institute, 2005).
46. Quoted in Connor, "Michel Serres: The Hard and the Soft."
47. Main, *Heartland.*
48. W. E. H. Stanner, *White Man Got No Dreaming* (Canberra: ANU Press, 1979).
49. Main, *Heartland.*

BIBLIOGRAPHY

Campbell, Hardy. *Campbell's 1907 Soil Culture Manual: A Complete Guide to Scientific Agriculture as Adapted to the Arid Regions*. Lincoln, NE: The Campbell Soil Culture Company Inc., 1907. http://archive.org/details/campbellssoilcul00camprich.

Connor, Steven. "Michel Serres: The Hard and the Soft" (lecture). Centre for Modern Studies: University of York, 2009. http://www.stevenconnor.com/hardsoft/.

Deleuze, Gilles and Felix Guattari. *A Thousand Plateaus: Capitalism and Schizophrenia*. Minneapolis: University of Minnesota Press, 1987.

Egan, Timothy. *The Worst Hard Time: The Untold Story of those Who Survived the Great American Dustbowl*. Boston: Houghton Mifflin Harcourt, 2006.

Farrell, Michael. "A Poetics of the Plough: Ned Kelly's *Jerilderie Letter*." *Southerly* 72:2 (2012), 77–96.

Food and Agricultural Organisation of the United Nations (FAO). "Drylands People and Land Use." *Waters and Cereals in Drylands*. London: Earthscan, 2008. http://www.fao.org/docrep/012/i0372e/i0372e01.pdf.

Grieves, Victoria. *Aboriginal Spirituality: Aboriginal Philosophy: The Basis of Aboriginal Social and Emotional Wellbeing* (discussion Paper no. 9). Darwin: Cooperative Research Centre for Aboriginal Health, 2009. http://www.crcah.org.au/publications/DP_series.html.

———. "Narranga Barrangang: Self and History, a Contemporary Aboriginal Journey." In *Ngarpartji Ngarpartji: Ego-Histoire from Europe and Indigenous Australia*. Eds. Cole, Haag, Castejon and Hughes. Canberra: ANU ePress, 2014.

Hohaia, Te Miringa, Gregory O'Brien, and Lara Stringman. *Parihaka: The Art of Passive Resistance*. Wellington: Victoria University Press, 2006.

Ihimaera, Witi. *The Parihaka Woman*. London: Vintage, 2011.

Jones, Alan. "A Brief History of the Plough." *Society of Ploughmen* (n.d.). http://www.ploughmen.co.uk/about-us/history-of-the-plough.

Main, George. *Heartland: The Regeneration of Rural Place*. Sydney: University of New South Wales Press, 2005.

March, George Perkins. *The Earth as Modified by Human Action*. New York: Charles Scribner, 1865.

Millennium Ecosystem Assessment. *Ecosystems and Human Well-being: Desertification Synthesis*. Washington, DC: World Resources Institute, 2005.

Moore, Laurie and Stephan Williams. *The True Story of Jimmy Governor*. Crows Nest: Allen and Unwin, 2001.

Muir, Cameron. *The Broken Promise of Agricultural Progress: An Environmental History*. London: Routledge Environmental Humanities Series, 2014.

Olson, J. C., J. J. Montag, and R. C. Nagle. *History of Nebraska*. Lincoln: Nebraska University Press, 2015.

Serres, Michel. *The Five Senses: A Philosophy of Mingled Bodies (I)*. Trans. Margaret Sankey and Peter Cowley. London: Continuum, 2008.

———. *The Natural Contract*. Trans. Elizabeth Macarthur and William Paulson. Ann Arbor: University of Michigan Press, 1995.

———. *The Parasite*. Trans. Lawrence R. Schehr. Baltimore: John Hopkins Press, 1982.

Stanner, W. E. H. *White Man Got No Dreaming*. Canberra: ANU Press, 1979.

Stromberg, Joseph. "What is the Anthropocene and Are We in It." *Smithsonian Magazine*, (January 2013). http://www.smithsonianmag.com/science-nature/what-is-the-anthropocene-and-are-we-in-it-164801414/?no-ist.

Vaughan, Adam. "Human Impact Has Pushed Earth into the Anthropocene, Scientists Say." *The Guardian*, January 7, 2016.

Wilber, Charles Dana. *The Great Valleys and Prairies of Nebraska and the Northwest*. Omaha, NE: Daily Republican Printing Company, 1881.

Wood, Marilyn. "The Breelong Blacks." *Race Matters: Indigenous Australians and "Our" Society*. Ed. G. Cowlishaw and B. Morris. Canberra: Aboriginal Studies Press, 1997.

Chapter 6

Conserving Land through Kindly Use and Reciprocity

Using the Land and Being Used by the Land

Anne O'Brien

Many problems of the Anthropocene are reproduced and carried forth through conventions of modern land use. Institutions of industrial agriculture and landscape management are dominated by a narrow conception of use, centred on satisfying demands that are produced in and through markets. Environmentally destructive consequences are only addressed to the extent that is strictly required by law or by the production process itself. In contrast, more ancient, time-tested practices of Australian Aboriginal use enshrine reciprocal obligations to the land in culture, ontologies and ritual. In the absence of cosmologies that recognize land as inhabited by beings towards which we have duties, members of non-indigenous cultures need other ways of foregrounding interdependence, developing ethical imaginations to *broaden and deepen* our concept of use. In conventional forms of manual and machine-mediated labour with the earth, practices of care towards the land are possible, but they often require considerable latitude from narrow, exclusively instrumental, control-oriented standard production protocols to allow for responsiveness and concern for the nonhumans involved in the labouring process. Transforming practice at least requires recognition of the nonhumans that enable our use of the land, while also noting and enabling their uses of the land, their vulnerability and needs. In this chapter, I use Wendell Berry's concept "kindly use" as well as Aboriginal concepts of use to consider the conditions that support deeper and broader conceptions of use and intimate, reciprocal and receptive relationships of land stewardship. I analyse how practitioners extend care towards the generative sources of use, particularly land and soil, using two fieldwork case studies of near-monoculture production in which land stewards learn to sustain reciprocal relationships with the land they work upon, particularly developing skills to care for the microbial ecologies of the soil.

BACKGROUND FOR THE INVESTIGATION

I entered adulthood during the Millennium Drought: a period of rural crisis throughout Eastern Australia lasting nine years. Images of despondent farmers tending to skeletal sheep, staring out towards desolate horizons were common on TV and in newspapers. Industrial agriculture appeared very tenuous amidst the dry. For most Australians on the tiny East coast strip of green, the magnitude of suffering was far less than those in rural areas who depend on regular rain for their livelihoods. Our sports fields became hard and dry, bushfires intensified and water restrictions reduced garden water use to twice a week, outside sunlit hours. As a gardener, I struggled to grow plants that had been domesticated in other parts of the world, using techniques that also were also from elsewhere: those plants became shriveled husks. All this left me with a strong hunch that European immigrants have never been quite "at home" in the Australian landscape.

As a non-Aboriginal Australian descended from mostly Irish immigrants who settled in rural New South Wales and Queensland from the 1840s onwards, the story of agriculture's long-term decline due to land degradation as well as economic consolidation in Australia can be read in my own family history. While virtually all my family members now live in major cities such as Sydney, tracing the locations of great grandparents' homes evokes an earlier time in which rural employment was easy to find, in places such as Brewarrina, Wellington, Araluen, Pambula, Bathurst and coastal Queensland, when the richness of the land built by Aboriginal traditional practices of use had not been depleted to the extent it has now. My mum's childhood summers with her cousins were spent on the Hay plain along the Murrumbidgee River. She describes the tree stumps that extended as far as the eye could see on the dry soil surface. The loss of this tree cover, combined with overgrazing, intensive use of ploughs and predominantly annual cropping increased the impact of salinity and degraded the soil: particularly reducing levels of organic matter. This in turn decreased resilience to drought, which amplified the harshness of the heat and the dryness of the climate, and was only alleviated by irrigation. Today, the main river system of Australia's food bowl, the Murray–Darling has a toxic blue-green algal bloom that runs for around 500 kilometres,[1] and the Darling River has run dry in parts of New South Wales for the third time in twelve years.[2]

THE PROBLEMS WITH MODERN CONCEPTS OF LAND USE

"Use" is popularly understood as both a generic verb and a pejorative, and its pejorative meaning bleeds into its generic meaning. When a human feels "used," she feels treated as only the means to an end, without adequate regard for her own ends. The grievance arises from a deficit of recognition, interaction, sensitivity and "give and take." The modern, *narrow* conception of use institutionalizes an alienated and extractive relationship to the world that renders nonhuman objects as of instrumental value alone, replaceable means to anthropocentric ends rather than both means and ends in themselves. This is a one-way interaction in which one dominant agent exercises control over the object of their production or consumption. When the implied goals are profit maximization and/or efficiency, it is often seen as a distraction to care for the qualities of the environment that make land use possible. In this mode, the only external feedback taken into account is that which enhances the task at hand. Thus, this narrow form of use is effectively closed to attitudes of care, curiosity or aesthetic appreciation. Environmental degradation often follows as a consequence.

An exclusively instrumental, narrow stance of use is often dictated by the constraints of modern production chains, as well as technologies that predetermine patterns of action and intervention.[3] It is maintained by coercion in the workplace, hierarchies that determine the temporalities and objectives of work without regard for the space and time required for an understanding of context in order for practitioners to develop good judgment and a working relationship with ecology. It is further compounded by the division of labour, by technological automation, and bureaucracy that distances managers and workers from the effects of a production process. Workers and managers assume a shallow, reifying stance as their default pattern, switching off capacities for care, thought and critical judgement towards the ecological consequences of their actions.

The Wise Use "movement" was launched in the United States almost thirty years ago in order to advance the agendas of logging and mining corporations in alliance with hunting and motoring enthusiasts seeking greater access to public lands.[4] A key plank of their discursive strategy in battling the environment movement was to monopolize the concept of use, and thus to frame environmentalists as impractical. This strategy also sought to drive a wedge between environmentalists and working-class communities built primarily around employment from extractive and agricultural industries. This frame falsely implies that the only way to use land productively is through a narrow form of use: to maximize control for maximum output, without regard for ecological well-being or for future generations. To criticize environmentalists as impractical in this context aligns profit-maximizing enterprise with practicality, reinforcing the belief that there is no alternative to exploitation.

If environmentalists respond by disavowing use and reinforcing the dichotomy between use and care, their access to the practical realm is discursively blocked. An avenue for advocacy is also lost when environmentalists consequently fail to engage with the innovations, joys and difficulties of manual labour: blaming land stewards for their role in environmental destruction. This reinforces the "jobs versus the environment" dichotomy, cutting off environmentalists from communities that depend on use of the land.

Instrumentality and *using the earth* as such are not inherently to blame for the environmental crisis in the Anthropocene. If use of the earth involves care as it does for Indigenous people, or for Wendell Berry in his concept *kindly use*, it can remind us of our interdependence with other species, particularly if our practices or technologies bring forth the Earth in such a way that its vitality, its diverse capabilities, its connectivity and its needs are more evident. Where instrumentality becomes a problem is where it narrowly and shallowly defines the relationship of use entirely: when there is a failure to treat the means of use as ends in themselves.

FIELDWORK ANALYSIS

My field research examined how land stewards are learning to care for the land through building soil ecological integrity and resilience. I am interested in the connections built through these practices, and how certain ecological objects come to matter for these people who learn to care for soil ecosystems while using them to produce certain commercial outputs or outcomes.

I visited Marie's sheep and wheat farm on the Western Australian wheat belt during a severe drought.[5] The orange dust blew in a *willy willy* through the landscape. Like many places in the world, soil ecologies in Western Australia have changed dramatically with the growth of industrial agriculture. David Murphey wrote in his book *Earthworms in Australia*, "As recently as 1979, around Albany in Western Australia farmers spoke of worms hanging on cultivator lines like spaghetti."[6] Yet, while the impact of industrialized land use has been more recent than most parts of the world, the scale of land degradation has been particularly widespread in South-West Western Australia, as dryland salinity, caused mainly by vegetation clearing, has scalded large parts of the wheat belt. In these regions, the persistence and regeneration of some original mallee bushland along the roadsides and around the margins of the wheat fields, along with the replanting of native grasses become all the more important.

In conventional agriculture when the soil functions according to plan, it is taken for granted within an instrumental relationship – "ready-to-hand" to use Heidegger's terminology[7] – and is commonly viewed as a *container* holding nutrients with an uncomplicated relationship to the growing plants. When Marie's soil did not function as expected, it became an object of interest, a *broken tool* – "present-at-hand," to invoke Heidegger again – foregrounded as an object of puzzlement and inquiry:

> We have high calcium soils here, and we tested the soil and the plant tissues. What the soil test was telling us was not what the tissue test was saying. The plant tissue came out lacking calcium – so there was some kind of absorption problem with the calcium not getting in.

Here, Marie recounts the moment she began to consider the possible role of microbes in the soil. Realizing that something in the soil was blocking absorption, she began learning about microbiology and became convinced of the causal role of microbes in the disjuncture between calcium in the soil and calcium in the plant tissue. Heidegger argues that practices of use and of care are not mutually exclusive in an absolute sense, but rather in a temporal sense: at any one time, we are either using or appreciating. He describes a back and forth transition between having practical relations (ready to hand) and having a contemplative relation (present to hand). Objects "ready to hand" can suddenly appear as "present at hand," when tools break "in the conspicuousness of the unusable."[8] Suddenly that which was taken for granted in a narrow conception of use is seen in a new light, and revealed in its contingency: its vulnerability or agency. When practitioners run into problems or crises during the ordinary course of work, they are prompted to question things, and through this process of questioning the assumptions from which they previously proceeded no longer can be taken for granted. The practitioner enters into a process of searching, in which she becomes open to new forms of input, and develops a less instrumental relationship with the means of their work, which now becomes an end in itself, and an object of puzzlement and wonder: qualities have been revealed that are beyond human control that command respect.

Marie's curiosity led her on a process that culminated in seeing the soil anew as richly populated with microbial life. Soil microbes became significant for her. She undertook a "Soil Food Web" microscopy and composting training course, and attended other workshops with experts on soil microbiology. She dedicated herself to caring for soil life, unlocking the soil's capability to make nutrients such as calcium bioavailable to the plants and animals. She initially experimented with compost teas, analysing them under the microscope, as well as creating the simpler compost extracts, which she opted for in preference to compost teas as they are more predictable. She planted perennial native pastures, which ensured that the soil was fed yearlong with the sugars exuded by the roots, which in turn polymerize and accumulate carbon over time,[9] particularly through the action of mycorrhizal fungi. Minimizing disturbance of the soil surface, she uses low till techniques on the wheat fields, using a knife plough, and she uses small amounts of herbicide to keep weeds down, indicating that mechanical intervention would harm the land in this dry climate. The soil microbes were only the first piece of the puzzle. Soon Marie's attention was directed towards other aspects of the system. The sheep were key mediators, allowed to roam on the wheat stubble, in the bushland remnants and on the native grass pastures. Marie modified their diets, feeding them a concentrated nutrient mix to build up immunity before withdrawing worming treatment.

Organisms in the sheep droppings could now survive and thrive in the soil, helping developing soil structure and function. Marie points out that now the sheep have pink noses indicating their parasite-free status, thus they haven't suffered from the removal of worming treatment, partly because it is so dry here that parasites find it difficult to thrive anyway. Marie calls the sheep "the inoculators," claiming that due to their impact, you can now feel the springiness of the soil when you drive or walk over it. Even though the grip of the drought has worsened, the sheep are fatter than those on neighbouring properties. She has measured the carbon in the soil: levels have increased over the last three years, increasing the spongy texture and thus holding more water. Marie's relationship to the land is deepening in its reciprocity. Her new practices attune her more closely to the local context of the land upon which she lives and works. In this relationship, responsiveness matters: with feedback from organisms on the farm helping to modify practices. Thus, care for ecological robustness of the soil is now central to practices of use of the soil on Marie's farm.

My second example comes from the mid-north coast of New South Wales, a golf course that has developed a more reciprocal, caring relationship of use centred on the production and application of compost tea. Instead of the typical practice of using more chemicals than any agricultural operation to maintain perfect greens, this business now makes hot compost heaps and brews of aerated compost tea amended with seaweed and fish hydrolysate, spraying it on the greens and fairways, with promising results. They now use 90 percent less chemicals, which means better practitioner health, as the greenkeepers no longer must be tested yearly for health problems due to chemicals. They are nourishing and creating habitat for underground microbial ecosystems under the greens, building surprisingly long root systems and organic matter. Tom, the manager says, "We're able to apply a lot less water – I haven't watered for the last seven weeks … in years before, under the same weather conditions, we would have watered several times by now. Even on the fairways, there's pure sand under the turf, and it's starting to hold a lot more water."[10] Instead of spraying alligator weed with herbicide on the sensitive watercourses, they now work with the local catchment management group to encourage the proliferation of beetles that graze on the leaves. The local council has signed the golf course up as a supplier of beetles to help farmers in the region with integrated pest management.

In a demountable office is a microscope with its field of view displayed on a computer screen. Tom points to a blurry thread: "That's our first nematode there. I carried on like an idiot here when I saw it," he says proudly. If you know gardening and agriculture, you'll know how strange this sounds. Nematodes are usually seen as bad guys: they eat plant roots. This man and I studied together at the Soil Food Web Institute's course, which taught us that the nematodes only eat roots when other food webs have broken down, when their preferred food source, bacteria, are limited in number. Nematodes in the right circumstances are beneficial to soils. While Tom's initial environmental motivations were weak, his concern for soil organisms grew as a result of adopting new methods and tools that disclosed the diversity and efficacy of soil life, while also highlighting their vulnerability. Tom's initial motivation was instrumental: he adopted the new practices because he was afraid of losing his job. Skeptical about the "greenie" methods, he reluctantly came on board because the previous manager lost his job for disagreeing with the boss.[11] It was surprising to see how much this man developed a creative relationship with his work after such initial coercion. When I asked if he enjoyed his job more, he said, "Absolutely."

The rich structuring of interdependent relationships in the environment gains significance for practitioners, moving them to broaden the goals of land stewardship so that benefits can be enjoyed by these other species, such as habitat and forage areas: so one human form of land use creates "spin off" uses for other organisms. The signals given by cultivated and uncultivated species (however inchoate) are interpreted as feedback by land stewards, helping to increase sensitivity, modifying patterns of work where a clear message is grasped, and helping build responsive relationships. Repertoires of increased attentiveness to the soil are integrated into everyday practice, using disclosive tools and technologies that mitigate the distancing effect of labour-saving technologies. Whereas most technologies in agriculture and other land stewardship are intended to reduce labour costs, in the case of the microscope and similar monitoring technologies, the goal is gaining better understanding of the compost tea, which is applied to the land. The assemblage activated by this type of use makes land stewardship a more highly skilled craft, attending just as much to the well-being of the soil as to the texture of the green surface.

In both case studies, broader (more open-ended) and deeper (more caring and knowledgeable) forms of use are enabled. These are each dimensions of what Wendell Berry calls "kindly use."

KINDLY USE

In his 1977 classic *The Unsettling of America*, Wendell Berry advocates for *kindly use* that is intimate, sensitive and reflexive, attenuated to particular contexts but accommodating many provisional objectives. To enable freedom of action, Berry argues that use should be grounded in small economic units rather than large instutitions:

> Institutional solutions tend to narrow and simplify as they approach action. … Organizations tend to move toward single objectives – a ruling, a vote, a law – and they find it relatively simple to cohere under acronyms and slogans.
>
> But kindly use is a concept that of necessity broadens, becoming more complex and diverse, as it approaches action. The land is too various in its kinds, climates, conditions, declivities, aspects, and histories to conform to any generalized understanding or to prosper under generalized treatment. The use of land cannot be both general and kindly. … Kindly use depends upon intimate knowledge, the most sensitive responsiveness and responsibility.[12]

Kindly use is alert to the diverse needs that become evident only after long periods of observation and interpretation. By engaging in and modulating their use of the land, stewards can improve their knowledge and responsiveness to the needs of ecosystems. Kindly use requires practitioner engagement: both means and ends are charged with significance and intrinsic value, rather than being reified.[13]

Berry contests the epistemic commitments of mass production agriculture, arguing that kindly use must respond to the specific qualities of land in which one lives and works, rather than generalized universal maxims. Modern agriculture rests on representational knowledge that affirms the generalized and the standardized as more worthy of regard than knowledge that is manifest and grounded in specific places. Such knowledge frameworks counsel little practical benefit in relations of intimacy and complexity, instead striving towards abstract mechanical relations, in order to construct industrial monocultural production landscapes that can be managed through standardized automated and machine-based intervention. Berry constructs a human archetype of capitalist modernity that he calls "the vagrant sovereign" who "does not know where he is morally," and cannot observe limits, even those that to previous generations would have appeared as *natural* and unchanging: "he assumes there is nothing he *can* do that he should not do, nothing that he *can* use that he should not use."[14] The geographical separation of home and work forged by the division of labour in modern human life breaks apart a basic intuition generated from our animal condition, embodied in the inclination to refrain from befouling one's nest.[15] Berry thus argues for the reunification of home and work, and for care towards landscapes in which one lives and works, "knowing where one is" and being responsible for the care of the places that surround us.[16]

One reason why agriculture in this era has become disconnected from its surroundings and why care for the landscape is difficult is that some technologies are configured to create or require uniformity in land surface, species selection, soil consistency, thus the small and the diverse do not easily fit within these parameters. Tim Ingold argues that distancing is a paradigmatic quality of modern technology: "The concept of technology, at least in its contemporary Western usage, sets out to establish the epistemological conditions for society's control over nature by maximizing the distance between them."[17] In contrast, "through their tools and techniques hunter-gatherers strive to minimize this distance [between humans and nature], drawing nature into the nexus of social relations, or 'humanising' it."[18] The "drawing in" that hunter-gatherers practice "has as its object to establish the conditions not of control but of a kind of mutualism."[19] In describing this relation, Ingold argues that it constitutes a different ontology: "The constitutive quality of their world is not intersubjectivity but *interagentivity*."[20] This involves an interdependent emergence of use, with body parts and materials "brought into use" rather than "used": intentionality and functionality are seen not as pre-existing properties, but "immanent in the activity itself. ... Whatever practitioners do *to* things is grounded in an attentive, perceptual involvement *with* them, or in other words they watch and feel as they work."[21] Such use must be pleasurable enough and interesting enough not to deaden the senses, and might be animated by different sources, such as religious revelation, the satisfaction of human embodied desire, or the joy of surprise and discovery.

Land stewards' proximity to the land in everyday work exposes them to signals from the land that can be interpreted in increasingly skillful ways whenever an open-ended, *poïetic* and responsive relationship is established. Thus, some conditions for the transformation of land stewardship are emergent within and immanent to the activity of stewardship itself if bodies and tools are recognized as resources for sensitivity. These non-instrumental stances need to be given enough space, time and support to become agents of transformation in everyday work. Today, more and more of Australia's land use takes place from a distance: viewed from the cockpit of a light plane during crop dusting, or from a computer hooked up to machines tracked by satellites and directed through the input of coordinates. A trial project is currently giving Australian beef cattle farmers access to data collected by NASA satellites that track the cattle's movements in remote Northern Australia, in a system that replaces several roles of embodied mustering.[22] While this gives a good overview picture of the land and animal movements, it would be a challenge to establish a close reciprocal connection of care-infused use with the land in the absence of a closer physical relation. This is in line with a general problem that modern practices of use – particularly those that aim to achieve economies of scale – can subvert the basic and primordial forms of accountability that come with embodied proximity.

INDIGENOUS PRACTICES OF RESPONSIVENESS

While learning about the transformations in ontology that arise in farmers' and greenkeepers' increased care towards the soil, I kept being reminded of Aboriginal care for the land. Sheep, wheat and golf represent some of the industries that have most degraded land and destroyed habitat in Australia and that have most benefited from colonial land relations. Aboriginal belief systems that regulate use prevent a narrow, exclusively instrumental stance towards the land because of the kinship relations between humans and other species, which also give these relations of use their depth. These understandings and identifications strengthen intimacy and responsiveness, as they are embedded in meaningful cosmologies that define interspecies obligations and land relations. In Australian Aboriginal cultures, the use of land does not imply environmental damage because relationships of use take place in the context of reciprocity, as part of maintaining direct or close connections of care, with plenty of time in between harvest events. A Wiradjuri friend and colleague, Jennifer Newman, has explained to me that Country suffers if it is not used.[23] Using helps to strengthen relationships because it involves attending to Country.

Inverting the environmental meanings of the terms "wild," and "domesticated," Deborah Bird Rose recounts the response of a Ngarinman senior custodian to viewing a highly eroded gully on his country:

> He looked at it long and heavily before he said: "It's the wild. Just the wild." Daly went on to speak of quiet country – the country in which all the care of generations of people is evident to those who know how to see it. Quiet country stands in contrast to the wild: we were looking at a wilderness, man-made and cattle-made. This "wild" was a place where the life of the country was falling down into the gullies and washing away with the rains.[24]

This reveals a very different understanding of *the wild* than that celebrated by Western environmentalists. Daly is using *wild* to lament a lack of care in a similar way that this word is sometimes used to describe children who suffer parental neglect. The land is acting out, not of freedom or health but of distress and fragmentation. At this stage it would be very difficult to recuperate the losses that have already been sustained, thus Daly's sadness. Rose interprets the term "quiet" in another paper:

> "quiet" is a Kriol term. It means tame, domesticated, not dangerous, under control. The inescapable irony, and the hurt for many Aboriginal people, is this: the country Europeans would want to see as "untouched wilderness" is the country that Daly Pulkara and others regard as properly cared for.[25]

Relations between Aboriginal people and the land are seen here as domesticated, enabling a process of making homes together, rather than a relationship of domination. In these close relationships of care, traditional owners are responsive to the particular needs of the land.

Responsiveness towards a landscape over time strengthens recognition and respect for individual entities in that landscape and the uses to which they can be devoted. When such responsiveness is culturally anchored, it is reflected in language and ritual. In his book on indigenous research methodologies Shawn Wilson, a Cree scholar living in Australia, shows how in his language, the naming of objects is based on their use. He writes: "In the Cree language, the literal translation into English for a chair would be 'the thing that you sit on' and the literal translation for a pen would be 'something that you write with.' … Objects themselves are not named; rather what they might be used for is described."[26] Thus the Cree language reflects ontological relations infused with perceptual recognition of practical possibilities. Wilson elaborates on an indigenous method of learning which involves increasing the number of relationships you have with things in order to know them better: "The more relationships between yourself and the other thing, the more fully you can comprehend its form and the greater your understanding becomes. … So the methodology is simply the building of more relations. … Our axiology demands that we be accountable to these relations that we form."[27]

Rose sees a similar awareness of connectedness as the basis of respect in Australian Aboriginal practices of knowing the land: "Respect is a matter of … knowing the connections so that one knows the many contexts in which respect is due, and knowing how to look after things so that one can fulfill one's role in life."[28] Such respect for the particular role that each individual entity plays in maintaining the order of the ecosystem is generated through attentive observation and thought:

> [Respect] requires constant attentiveness to the value of something or someone, appreciation of the fact that that individual has its own contribution to make which is vital to the natural order. … None of this is possible without careful thought and observation, so that one can come to better understand what its function or contribution is, what conditions are needed to permit its continuation, how not to interfere with it, and how best to enable it.[29]

In a similar way, Laurie Anne Whitt and others discuss Diné rituals of gathering plants for medicine, which involve attentiveness to the particular individuals being cared for: "When you are collecting medicine healing herbs, you have to collect for the individual sick person. You make an offering to the plants in your prayers. … All these medicine plants have a specific song and prayer to go along with them."[30] For many Indigenous groups, obligations and knowledge relations are situated in place to such an extent that if a place is irreversibly changed, the knowledge associated with that place can be lost: "it was Mr. Sequoyah's belief that if the Valley were flooded, he would lose his knowledge of medicine."[31] Whitt et al. emphasize the modes by which knowledge arises intimately, *in situ*, in contrast to Western science: "Western science, western knowledge of the natural world, is representational. Indigenous science, indigenous knowledge of the natural world, is … presentational. Its continuation, its transmission, its possibility turn vitally upon the presence of the natural world, and on the kind of experiences that world offers."[32] Thus such connection to ancestral land must be maintained for the knowledges of respectful use to remain available to future generations of Aboriginal custodians.

CONCLUSION

How can the dominant narrow, shallow, exclusively instrumental ontology of use of the land be transformed to make room for something more? For settler Australians it is a major challenge to connect with land in a broader and deeper way, to simultaneously *use* the land and also *be used by* the land, making space for and enabling other species to also make their livelihoods there.[33] The desire to control, reinforced through modern patterns of management and certain technologies, has been a key part of the problem: there is much to be learned from taking a step back and being humble, observing the patterns of nature, learning from the disclosive powers of local knowledge, science and technologies such as microscopes, apprenticing ourselves to ecological farmers as well as instating the conditions by which Aboriginal relations to land through practices of use can be recuperated.

Kindly use requires the incorporation of practices of sensitivity and responsiveness into day-to-day work. In Australia and in other colonized countries, this responsiveness should take on an additional political dimension of becoming answerable to the ongoing demands of land justice that still are unmet,[34] such as the negotiation of long-overdue treaties in Australia and the proper recognition of land and sea rights to harvest the bounty of the land.[35] This is particularly brought into sharp relief by the common feature of the case studies of both Marie and Tom's regenerative work being enabled through the fact that the *owners* of the land and the capital supported a new mode of stewardship. Likewise, throughout his work, Wendell Berry places a strong emphasis on the stability that ownership of the land confers, a relation that makes agrarian thought and politics uncomfortably complicit in colonial land relations. Despite this, I still find much of value in this tradition that should be part of an ongoing conversation on social and environmental justice.

In this chapter, I have developed a concept of kindly use of the land to respond to the ecological crises caused by the dominant modern and Western capitalist paradigm of use. The exclusively instrumental, rigid orientation to maximizing profit is deflated and pluralized, with provisional goals able to be modified in response to ecological feedback. In my fieldwork examples, I have shown how transformations in key aspects of the Western concept of land use are being enacted, particularly through care for the soil in regenerative land stewardship, which broadens and deepens a practitioner's stance from exclusively instrumental reason to receptivity. This orientation to the land, in its capability to renew an engaged stance (thus avoiding reification), has some resonances with the richly meaningful relationships that Aboriginal people maintain with their land, through practices of use, which involve reciprocal obligations to particular places and the species that make homes there. In these receptive relations, abundance is built through strengthening interconnectedness, opening up relations, becoming answerable to others rather than seeking to control. The popular Australian gardening writer Jackie French provides an example to emulate: "Our garden feeds us, the birds, three wombats, two echidnas, a mob of visiting kangaroos, seven possums and about 40,000 other species."[36]

NOTES

1. Chris McLennan, "Water Crisis Worsens as Murray River Blue-Green Algae Spreads," *Weekly Times Now*, March 9, 2016. http://www.weeklytimesnow.com.au/agribusiness/water/water-crisis-worsens-as-murray-river-bluegreen-algae-spreads/news-story/f9f6c3c1b3989eecccd13982d0e46a1d.

2. Emma Brown, "Darling River Irrigators Begin to Pull Out Wine Grapes and Citrus Trees to Save Water," *ABC Rural*, January 7, 2016. http://www.abc.net.au/news/2016-01-07/darling-river-irrigators-begin-to-pull-our-permanent-plantings-/7071086.

3. I am following Charles Taylor's use of the term "exclusive humanism" to describe a world view that is closed to transcendental understandings of spirituality or religion. See Charles Taylor, *A Secular Age* (Cambridge: Harvard University Press, 2007), 19.

4. Sharon Beder, *Global Spin: The Corporate Assault on Environmentalism* (Carlton North, VIC: Scribe Publications, 2000), 47.

5. "Marie" is not the interviewee's real name.

6. David Murphy, *Earthworms in Australia* (South Melbourne: Hyland House, 1993), 29.

7. Martin Heidegger, *Being and Time* (1962), trans. John Macquarrie and Edward Robinson (New York: Harper and Perennial, 2008), 104.

8. Ibid.

9. See Tracy Frisch, "SOS: Save Our Soils: Dr. Christine Jones Explains the Life-Giving Link Between Carbon and Healthy Topsoil," *Acres USA* 45:3 (2015): 3–4. http://www.acresusa.com/media/downloads/March15_Jones.pdf.

10. Tom's name has also been changed.

11. Certain power relationships enable the changed relations of use. In the first case study, Marie's ownership of her land and the capital on it gives her leeway to experiment. In the second case study, Tom was pushed into a direction of regenerative land stewardship by the owner of the business. This reflects the continuing enmeshment between use and private property under capitalism.

12. Wendell Berry, *The Unsettling of America: Culture and Agriculture* (San Francisco: Sierra Club Books, 1977), 31.

13. According to Axel Honneth, reification is a "distorted, atrophied form of praxis," "in which one's natural surroundings, social environment and personal characteristics come to be apprehended in a detached and emotionless manner." From Axel Honneth, *Reification: A New Look at an Old Idea* (Oxford: Oxford University Press, 2008), 25–26.

14. Berry, *The Unsettling of America*, 53.

15. Ibid., 51.

16. Ibid., 53.

17. Tim Ingold, *The Perception of the Environment, Essays on Livelihood, Dwelling and Skill* (London: Routledge, 2000), 314.

18. Ibid.

19. Ibid.

20. Ibid., 47. Emphasis in original.

21. Ibid., 352–53.

22. Rachel Pannett, "NASA Helps Aussie Ranchers Herd Cattle," *Wall Street Journal Video*, January 24, 2016. http://www.wsj.com/video/nasa-helps-aussie-ranchers-herd-cattle/B7FC15E4-4965-4274-BA35-5279B89E920E.html.

23. Jennifer Newman, personal communication, March 2015.

24. Deborah Bird Rose, *Nourishing Terrains: Australian Aboriginal Views of Landscape and Wilderness* (Canberra, Australian Government: Australian Heritage Commission, 1996), 19.

25. Deborah Bird Rose, "Exploring an Aboriginal Land Ethic," *Meanjin* 4:3 (1988): 386.

26. Shawn Wilson, *Research is Ceremony: Indigenous Research Methods* (Halifax, CA: Fernwood Publishing, 2008), 73.

27. Ibid., 79.

28. Deborah Bird Rose, "The Ecological Humanities," in *Manifesto for Living in the Anthropocene*, ed. Katherine Gibson, Deborah Bird Rose, and Ruth Fincher (Brooklyn, NY: Punctum Books 2015), 5. Quote within the quote from Deborah Bird Rose, Diana James, and Christine Watson, *Indigenous Kinship with the Natural World in New South Wales* (Hurstville: NSW, National Parks and Wildlife Service, 2003), 67.

29. Laurie Anne Whitt, Mere Roberts, Waerete Norman, and Vicki Grieves, "Belonging to Land: Indigenous Knowledge Systems and the Natural World," *Oklahoma City University Law Review* 26 (2001): 728.

30. Affidavit of Mae Tso 35, at 328, Manybeads v. United States, 730 F. Supp. 1515 (D. Ariz. 1989) (No. 88–410), quoted in Whitt et al., "Belonging to Land," 727.

31. Whitt et al., "Belonging to Land," 702.

32. Ibid., 704.

33. For a fuller discussion of more-than-human livelihoods, see J. K. Gibson-Graham and Ethan Miller, "Economy as Ecological Livelihood," in *Manifesto for Living in the Anthropocene*, ed. Katherine Gibson, Deborah Bird Rose, and Ruth Fincher (New York: Punctum Books, 2015), 7–16.

34. For a discussion of receptivity as answerability, see Nikolas Kompridis, "Recognition and Receptivity: Forms of Normative Response in the Lives of the Animals We Are," *New Literary History* 44 (2013): 1–24.

35. Many Aboriginal people continue to be imprisoned on the NSW South Coast for gathering food such as Abalone, that they call "Mutton fish." See Bronwyn Adcock's report on the Australian Broadcasting Corporation's *Background Briefing*, "'Strangers in our own place': Aboriginal community fights for cultural fishing rights," February 18, 2016. http://www.abc.net.au/radionational/programs/backgroundbriefing/aboriginal-fishers-report-harrassment-fishing-traditional-food/7180326.

36. Jackie French, *The Wilderness Garden* (Melbourne: Aird Books, 1992), 2.

BIBLIOGRAPHY

Adcock, Bronwyn. "'Strangers in Our Own Place': Aboriginal Community Fights for Cultural Fishing Rights." *Background Briefing*. Australian Broadcast Corporation, 2016. http://www.abc.net.au/radionational/programs/backgroundbriefing/aboriginal-fishers-report-harrassment-fishing-traditional-food/7180326.

Beder, Sharon. *Global Spin: The Corporate Assault on Environmentalism*. Carlton North, VIC: Scribe Publications, 2000.

Berry, Wendell. *The Unsettling of America: Culture and Agriculture*. San Francisco: Sierra Club Books, 1977.

Brown, Emma. "Darling River Irrigators Begin to Pull Out Wine Grapes and Citrus Trees to Save Water." *ABC Rural*, January 7, 2016. http://www.abc.net.au/news/2016-01-07/darling-river-irrigators-begin-to-pull-our-permanent-plantings-/7071086.

French, Jackie. *The Wilderness Garden*. Melbourne: Aird Books, 1992.

Frisch, Tracy. "SOS: Save Our Soils: Dr. Christine Jones Explains the Life-Giving Link Between Carbon and Healthy Topsoil." *Acres USA* 45:3 (2015): 3–4. http://www.acresusa.com/media/downloads/March15_Jones.pdf.

Gibson-Graham, J. K. and Ethan Miller. "Economy as Ecological Livelihood." *Manifesto for Living in the Anthropocene*. Ed. Katherine Gibson, Deborah Bird Rose and Ruth Fincher. New York: Punctum Books, 2015.

Heidegger, Martin. *Being and Time* (1962). Trans. John Macquarrie and Edward Robinson. New York: Harper and Perennial, 2008.

Honneth, Axel. *Reification: A New Look at an Old Idea*. Oxford: Oxford University Press, 2008.

Ingold, Tim. *The Perception of the Environment, Essays on Livelihood, Dwelling and Skill*. London: Routledge, 2000.

Kompridis, Nikolas. "Recognition and Receptivity: Forms of Normative Response in the Lives of the Animals We Are." *New Literary History* 44 (2013): 1–24.

McLennan, Chris. "Water Crisis Worsens as Murray River Blue-Green Algae Spreads." *Weekly Times Now*, March 9, 2016. http://www.weeklytimesnow.com.au/agribusiness/water/water-crisis-worsens-as-murray-river-bluegreen-algae-spreads/news-story/f9f6c3c1b3989eecccd13982d0e46a1d.

Murphy, David. *Earthworms in Australia*. South Melbourne: Hyland House, 1993.

Newman, Jennifer. Personal Communication, March 2015.

Pannett, Rachel. "NASA Helps Aussie Ranchers Herd Cattle." *Wall Street Journal Video*, January 24, 2016. http://www.wsj.com/video/nasa-helps-aussie-ranchers-herd-cattle/B7FC15E4-4965-4274-BA35-5279B89E920E.html.

Rose, Deborah Bird. "Exploring an Aboriginal Land Ethic." *Meanjin* 4:3 (1988): 378–87.

———. *Nourishing Terrains: Australian Aboriginal Views of Landscape and Wilderness*. Canberra: Australian Heritage Commission, 1996.

———. "The Ecological Humanities." *Manifesto for Living in the Anthropocene*. Ed. Katherine Gibson, Deborah Bird Rose, and Ruth Fincher. Brooklyn, NY: Punctum Books, 2015.

Rose, Deborah Bird, Diana James, and Christine Watson. *Indigenous Kinship with the Natural World in New South Wales*. Hurstville: NSW, National Parks and Wildlife Service, 2003.

Taylor, Charles. *A Secular Age*. Cambridge: Harvard University Press, 2007.

Whitt, Laurie Anne, Mere Roberts, Waerete Norman, and Vicki Grieves. "Belonging to Land: Indigenous Knowledge Systems and the Natural World." *Oklahoma City University Law Review* 26 (2001): 701–43.

Wilson, Shawn. *Research is Ceremony: Indigenous Research Methods*. Halifax, CA: Fernwood Publishing, 2008.

Chapter 7

"One Loaf of Bread at a Time"

System Change through Community Food Initiatives

Sharon Stevens

Eating. Seed transforms sunshine, air, water and soil into complex plant forms. These are again transformed, by eating, into our animal bodies. We eat, and we become inextricably bound, symbolically and materially, within consequential relationships with our environment.

Our interconnectedness with other life does not, however, imply either harmony or symbiosis, and our dominant modes of sourcing food in the Anthropocene are among the most damaging participants in the rapid socio-ecological change indicated by that term. Our eating has become out of sync with time and space, and with the seasonal foods of our local environment: our nutrition and health decline as a result.[1] Our communities decline as well. Hapū or settlement gardens once helped sustain – and were in turn sustained by – the interconnection of community life and livelihood.[2] Gardening know-how was once a cultural inheritance, but there is now a generation gap, with knowledge lost like heritage plants. We have forgotten how to save seed, how to plant, how to harvest, how to share, how to cook, how to eat from the earth.[3] Instead, we eat from the supermarket, and from the global food system behind it.

This is costly. It is costly for people, with the logistics of oil-hungry transportation and food storage together raising prices enough that those on constrained budgets turn to processed over fresh foods.[4] It is costly for the earth and its species, with even the habitat necessary to home our pollinators destroyed by increasingly large-scale commercial agribusiness.[5] It is costly for farmers, who can't afford better practices, who plough under food that isn't ripe on the day set aside for harvesting.[6]

From earth we have come, and to earth we will return – but our food is losing its earthiness, our daily bread its connection with soil. Some of us eat so many preservatives that even our dying bodies no longer know how to decompose with dignity. If we are what we eat, then we are monocropping, chemical pest control, artificial fertilization, climate change, soil toxicity build-up, biodiversity loss and resource scarcity. We are GMOs; we are the abrupt rupture between our food and the slow and symbiotic relations of mutual adaptation within cultures of place. We are massive forest-to-farm conversions and tillage practices that expose soil to wind, sun and erosion, that destroy whole watersheds and habitats, and that rapidly release excessive amounts of carbon into the atmosphere. We are neoliberalism: growth economies, market expansion, the concentration of wealth, short-term profits, efficiency, productivity. From culture to culture, eating has been understood as communion. We are alienated exchange.

Our alienation has many harmful effects. Among them is how our perceived separation from nature obscures available choices for responding within (taking responsibility for) the particularities of our entanglements with the rest of our world. At the grass-roots, however, there is a groundswell of people working to realign how eating connects us within our local socioecologies and, by extension, within our planetary system. These are people who farm, who garden, who cook and who mentor. They reconnect growers and eaters, highlighting the inseparability between healthy soil and healthy humans, while raising awareness, passing on skills and actively living out their confidence in the agency of grassroots communities. The actions taken by these grassroots organizers are an ethical response to the conundrums of our age. Their actions assert that we can co-create, in synergy with other aspects of our planetary system, a more self-enhancing co-evolutionary pathway.

This chapter is a form of "listening" to grass-roots actors. Those interviewed are people who have chosen to stick to community-based organizing in spite of regular deterrents. What sustains these actors in the face of despair-sized global problems? A shared critique of globalized industrial agriculture is insufficient to explain their motivations, their hope. Why do they believe their actions can manifest a viable alternative to the hegemonic and destructive entanglements that dominate our current practices of eating? How do these grass-roots actors theorize their aspirations and their experiences?

Moreover, what do those who privilege direct action have to offer within discursive contexts that are traditionally set aside for language-centred forms of interdisciplinary theory-making? What those I interviewed make apparent is that the initiatives in which they participate are not "just" about the next meal or ecologically sounder approaches to growing food. These may be essential components, but their grass-roots leaders stress that their initiatives are pre-eminently about social relationships, about food and nutrition and about initiating systemic socioecological transformation through "slow and small solutions."[7] These initiatives are about eating in order to reconfigure how we participate in entanglements. In the words of one organizer, "If you eat right in a daily, inconsequential, unremarkable way, you're connected and doing something that's good to do, with profound implications on many levels."[8] In short, these organizers understand their practical actions to be the most appropriate response they can make to the ethical imperatives of the Anthropocene.

The community food initiatives in which these grass-roots actors are involved can, nonetheless, seem inconsequential – or at least mild, small and slow – when compared to other forms of environmental action. The patience of gardening is a stark contrast with the image of environmentalism as protest politics that developed in the 1970s and 1980s. Many of those I interviewed are engaged from time to time in forms of non-violent protest, largely because community change-agents are among those who directly experience how the policies of councils and governments can make it more or less easy to co-create a positive future. But if we have correctly defined the problems with food systems to be a consequence of globalization, then their solutions must be local; if they are a consequence of an inherently growth-oriented money system, then their solutions must be some other approach (or some other combination of approaches) to sharing (exchange) and to reconnecting people around growing and eating.

That is, our society and planet need something different than the hegemonic global food systems that currently dominate our day-to-day consumption. Marketplace consumption focuses attention on transaction over connection, thereby creating an illusion of separation and a reality of alienation that make ethical response options more difficult to feel, to identify or to act upon. Those who take direct action at the grass roots are committed to the step-by-step labour of building an alternative, a system where growing food is once again place-based and small-scale, where solutions must consequently also be local in their implementation, although also synergistic and mutually reinforcing across the globe. Actions taken at non-local scales (e.g. global trade policies, national regulatory frameworks governing food handling, regional plans and their impact on land availability) can either enable or restrict local opportunities. In the final analysis, however, if the end we seek is a reconstructed system where more people are growing more food closer to home, then one of the most transformative things we can do is simply to embed ourselves consciously in the work of co-creating that relocalized system, discovering what we can learn by doing so. Our goal, in short, is to "detach ourselves from the destructive systems of neoliberal capitalism … one loaf of bread at a time."[9]

A FEW WORDS ON STANDPOINT

This chapter is based on interviews with a range of community organizers who, together with their collaborators, are putting considerable time, creativity and energy into community-based food initiatives. I am writing as a community organizer myself, primarily in my role as an employee of Environment Network Manawatu (ENM). ENM is a charitable umbrella organization that provides networking opportunities, publicity, advice and other forms of support for the activities of nearly fifty member groups (primarily but not exclusively NGOs) who are operating in the general region surrounding the watershed of Te Awa o Manawatū in central Te Ika a Māui o Aotearoa (the Manawatu River watershed in the central North Island of New Zealand), and who have agreed to jointly "promote a network of collective action to realise a vision of an enhanced Manawatu environment."[10] Network activities are directed at biodiversity and conservation, water quality, education, sustainable transport, resource use and recovery, and other environmental issues. My responsibilities include supporting and raising the profile of the community food initiatives that are being developed by a few more than 20 percent of the groups within this network. Outside of my ENM staff role, I work and volunteer in various food-related eco-initiatives.

Collectively, ENM's food-oriented member groups are involved in growing (school and community gardens and orchards, farming and mahinga kai [food-gathering] projects), distribution (volunteer picking networks, cooperative farm food box schemes and farmers' markets) and education and mentoring (in gardening, cooking, preserving and beekeeping, with various programmes directed at children, youth and adults). The work of these groups and the words of their representatives are the basis of this chapter. I reference those I interview by name, with their permission and that of the organizations for whom they speak. They and their organizations have additionally had opportunities to review interview notes and a penultimate draft of this chapter.

I dedicate the rest of this chapter to the claims of those I interview, so that my text reads, in large part, like a practitioner account. This raises a question of fit within an academic collection. In what way(s) can the self-understandings of those who take action enhance our theoretical understanding of the entangled relationships that cross – and call into question – boundaries between humans and nonhumans? This question is fundamentally epistemological, and it encompasses embedded questions about what theory looks like, who holds it, who enacts it and how it is developed. Do individuals make theory? Is it instead made within shifting intra-, inter-, and transdisciplinary relationships, and held within decentred, dynamic dialogues? Perhaps theory is also made in collectives that do not (choose to) speak by means of academic discourses, or perhaps even in collectives that blur the boundary between humans and the rest of the world? Consider: Who knows better than a traditional seed saver the interconnectedness of our cultivated food crops, their wild relatives (via continual gene transfer and renewal), particular very local ecologies (soil types, insects and their habitats, rainfall patterns, micro-climates), our cultural practices, our livelihoods and our selves? Who knows better the co-evolutionary transformation of all these elements?[11] When contemporary social actors work to consciously realign our relationships to soil, seed and community, are they not also creating theories about how we are impacted by our entanglements, and about what an ethical response might look like?

In any collective cultural practice, there is a way in which a theory is being actively articulated. Collective action makes a claim about how to orient ourselves ethically within our world, and it offers an explanation for an evolving reality, thereby guiding further action. To honour this, and at the risk of not making everything I hope to communicate sufficiently clear within an academic context, within this chapter I choose to use only a light touch when framing and translating the language of those I interview. While I wish to help build a bridge between those I interview and the anticipated (academic) readers of this book, I prefer not to foreground academic discourse. Instead, I stay closer to the language of those whose direct action embeds them deeply and strategically in relationships with food, relationships that blur boundaries and reconfigure how we relate to seed, soil, climate and one another.

My preference for the language of those I interview is informed by feminist standpoint theory.[12] In part, I write out of a concern with the way keeping identities confidential can contribute to the appropriation of community knowledge. More substantively, I wish to minimize the objectification of community participants that can occur in the process of translating their own language into that of academics, especially if that translation implies that participant language is somehow inadequate and requires supplementation. Most essentially, however, I write from an epistemological stance that highlights how that which has been marginalized (within any given context) has much to contribute. By foregrounding, within an academic collection, the accounts of community organizers, I am attempting to foreground an ethical responsiveness that reaches its highest expression through direct action. My attempt is limited by the fact that, in the end, this chapter is indeed a selective representation (in writing) of others' talk about their action, and it lacks the immediacy of understanding I experience when, day after day, year after year, I place myself in the garden alongside those I have interviewed, alongside pollinators, alongside small plants finding their place to emerge from the soil – all of us building community together. By staying as close to the action, to the earth, and to this broadly defined community as I know how to do within a print text, I seek to underscore our need to give priority to direct acts of co-creating alternative co-evolutionary pathways, pathways that are consciously chosen because they are better suited to sustaining our own (and many other) species, pathways that require us to more deeply embed ourselves within very local and earthy relationships as a challenge to the hegemonic and alienating practices of global markets and industrialized agriculture.

In selecting which aspects of organizer accounts to include, therefore, my focus is on how community action can manifest an alternative to the hegemonic food system.[13] The chapter is organized around three overarching themes: (1) how community actions can raise awareness, educate and contribute to sustainable, place-based food cultures; (2) how actions can participate in the development of relocalizing structures, spaces and systems; and, as the chapter conclusion, (3) how activists theorize the importance and effectiveness of their attempts to reconfigure, through eating, our entanglement with our local and planetary environment. This third theme, put more simply, is about why those I interviewed have so much hope that it influences their daily actions and, in many cases, their entire lives.

KNOWLEDGE, SKILLS, CONSCIOUSNESS-RAISING, AND WISDOM FOR THE REGENERATION OF GARDEN CULTURES

> Industrial agriculture is the biggest planet-wrecking activity there is. But people have to eat. ... Doing the community work is the positive approach. Instead of saying "don't do this," you say, "Do do this."[14]

In the upper Rangitikei, in the small town of Rangiwahia, REACT – The Rangiwahia Arts Centre Trust – is developing a rural centre to support workshops in art and sustainable resource use. The centre is also a hub of high-altitude and diverse food growing, and it offers an alternative approach to growing within a conventional farm community. REACT founders have found that being the change isn't always easy, but they have created a haven for the subgroup of Rangiwahia villagers who seek an eco-conscious approach to farming and growing in order to thrive. The REACT haven extends to other species. When a farmer-neighbour rescued heritage apple trees off a burn pile, REACT replanted the uprooted trees. This year the rescued trees have offered a bumper harvest.[15]

In the upper Manawatū River Catchment, on the eastern side of the Tararuas and on down through Te Āpiti, the Gorge, Te Kāuru Eastern Manawatū Hapū Collective are participating in a multiagency Manawatū River restoration project. They bring to their participation a mahinga kai framework involving connections between place and traditional foods, and they are drawing on their ancestral heritage to guide the revival of the culturally significant Parahaki Island. Clean-up and restoration are well under way, as is planning for a māra kai, or hapū garden, which is intended to complement the mature karaka grove on the riverbank across from the island. The aging grove stands witness to the role that food plantings once had along major transportation routes, with river pathways shaping, and being shaped by, human movements. Hapū elder and leader Manahi Paewai notes that community and place are inseparable. He sees the grove's continuity as an opportunity to redevelop a sense of identity: the grove links hapū to a history where they were "nourished by Papatūānuku [loosely translates as Mother Earth] ... as part and parcel of the communal system."[16]

In networks that stretch throughout the Manawatu and sometimes beyond, SuperGrans volunteer mentors work one-on-one in homes, and in occasional group workshops, teaching families skills that include cooking, gardening and preserving.[17] At Palmerston North Girls' High, the activities of the lunchtime Enviro Group include planting and maintaining their school garden; their plans for the 2016 school year included relocating and redeveloping the garden in a central location for greater consciousness-raising among their fellow students.[18] In Ashhurst, the sustainability education courses of RECAP are drawing participants not only from Palmerston North, the largest nearby urban population, but from both sides of the Tararuas; many of these practical workshops are based in volunteer-developed and -maintained community food plantings, a garden and an orchard, which provide free food to all. All these educational activities are based in conscious acts of reconnection intended to reconfigure how we access our food.

The ways we access our food – our food systems – are inextricably entangled with our cultural practices of consumption. Throughout this section, I focus primarily on actions intended to shift cultural practice, drawing on the comments of community organizers to illustrate how local food initiatives are designed to raise consciousness, bring environmentalism to new audiences, share and develop skills, serve as experimental spaces for cultural (re)-generation and co-create reconfigured relationships with the earth. In the next section of my chapter, I turn to systems change and the more structural dimensions of community organizing. This is an imperfect division, adopted for analytic purposes.

That said, the dimension of organizing that is explicitly oriented towards cultural change helps to underscore the ways that local initiatives can have greater impact than might at first be apparent. In the introduction to this collection, Sy Taffel and Nicholas Holm write that it is the "sum total of those heterogenous actions taken at local levels that have given rise to the Anthropocene. [... W]hen we seek to mobilize, alter, affect, impact or otherwise engage with the consequences of contemporary ecological crises, we can only do so as embodied actants ... taking action requires us to return to the here and now." Those interviewed for this chapter would readily agree. All of the organizers I spoke to were conscious of the centrality of local action, yet they thought strategically about how they might contribute to greater changes: that is, they designed their projects to have both immediate impact and ripple effects. For example, because of their desire for social outreach, the gardeners I interviewed prized both high traffic public spaces for community gardens and high public visibility for private front-yard gardens.[19]

Yvonne Marsh of Green Hub, who has toured many gardens on the South Island of Aotearoa, wants to scale up these types of activities through the development of a commercial yet still community-based market garden. Many who would stop by a visible garden along a well-used walkway or cycleway, Marsh tells me, would never come to a building-based eco-centre. Gardens can therefore serve as sites for a new environmental consciousness, and they would be ideal for integrating drop-in public workshops along with vocational skills training, spray-free food production for those seeking a healthier diet, and much more.[20] RECAP's activities incorporate many aspects of this education-oriented vision on the scale of a home garden. Transition Feilding finds that their public gardening facilitates all sorts of eco-conversations: about energy use, for example, or about how gardening sequesters carbon.[21] When people are already doing something positive with the soil, bringing up climate change is taking "a big, global problem back on a personal, practical level," Vanessa Witt of Transition Feilding tells me, getting people to engage what is "potentially a big scary issue" in a non-threatening way, within an immediate context of personal action that lends itself to further action.[22]

Other types of network food initiatives similarly yoke direct education and more general consciousness raising. Paul Jenkin of the Manawatu Beekeepers' Club recounts that, when he became Chairperson, the club chose to focus on supporting beginner beekeepers, on the one hand, while also running festival stalls not only to advertise the club but also to make people more aware of how to cultivate bee habitat and forage within their own home spaces.[23] Antonia McGhie, the student president of Massey University Sustainability Club (MUSC), is clear about her desire to share a type of eco-awareness throughout the entire student population. One of her first acts as president was to regularize an on-campus, student-oriented farmers' market, which the club prizes as a way to not only increase student access to fresh food, but also to increase student engagement and ensure students are "educated in the things that matter." McGhie is determined to raise awareness among her fellow students about the impacts of eating: bringing students closer to growers aids her aims.[24] She also wants students to experience a sense of community and a less alienated way of buying food.

For some, the deliberate transfer of skills for sustainability is at the heart of their missions. SuperGrans is an obvious example, with a core mission of providing mentoring for interpersonal (often intergenerational) skills transfer for "safe, healthy, low-impact, budget living."[25] Plant to Plate takes another route to skills sharing, with its emphasis on primary school gardening and cooking workshops. "Gardening is a forgotten art," Plant to Plate Coordinator Cindy Duker tells me, a comment echoed by others who recall the school-based growing competitions of their childhood[26] or gardening alongside their families: "Vanessa's first bought present was a spade," says Ellen Witt of Transition Feilding about her daughter.[27]

It is apparent to many on the ground that local initiatives are just beginning to address the community hunger for knowledge. RECAP's subsidized workshops are fully subscribed more often than not; both Plant to Plate and Supergrans find that the amount of classroom teaching and home mentoring that they respectively undertake is regularly limited by the availability of skilled gardening volunteers, which is insufficient to meet client demand.[28] "We need to relearn the skills of our elders before that knowledge is lost."[29] Where is this community-building around food heading? Not back to the past, according to Harvey Jones of RECAP: first, we're "not carving a living out of the bush" and, second, with increased access to information and technologies, we have an opportunity to "work smarter."[30] Still, change will be a challenge. Growing food requires a change in living patterns because vegetables need watering in January no matter how busy people get. People won't be willing to return to historic, pre-mechanized farming patterns where they had no choice but to work on the farm or perhaps in the freezing works (the national term for what Europeans call an abattoir).[31] Living in a society with more choice, if we want to reduce mechanization and increase direct engagement, we'll need to develop new approaches, new social relations and a new culture of food production.

This may be one of the more important cultural contributions of community food initiatives, from the prototypical small farm of REACT to Te Kāuru's commitment to take the best of the past and bring it forward to the future. Our current socioecological relationships are headed towards poor outcomes for humans and many other species, and our past relationships cannot be recreated. Those who are taking practical action to find and share new and more conscious forms of livelihood are thereby working to shift the processes and direction of our co-evolution within the planetary system.

In the age of the Anthropocene, where our current social relations are having unprecedented impact, the work of transforming human cultures may very well be the most challenging aspect of the shifts required for humans to overcome alienation and adopt a less self-destructive pathway. As Kim Penny of SuperGrans tells me, cultural experimentation and renewal are going to require "a level of openness and honesty … wisdom, maturity, and cold hard enthusiasm."[32] Maybe some people will get on with the work of just making things happen, but you still need someone with the wisdom and people care skills to put new social structures in place, to think about how community works and to help develop ground rules for cooperation: "You need someone wise enough to help the community grow while the veggies are growing."[33]

LOCAL FOOD NETWORKS AND ACTS OF SOCIAL RECONNECTION

> "The idea of what we're doing is no longer strange and new. Every time we get the idea out there, it becomes normal to share. When people start [sharing fruit and other produce] more independently because of the consciousness-raising role of having a group like Community Fruit Harvest out there, then you see social change on a broader level."[34]

Neoliberal economic policies have left those who wish to relocalize with a Catch 22[35] that arises from a complex of social alienation, wealth accumulation and poverty, degraded environments, and deteriorating place-based cultures. In the case of food, for local growers to be successful requires ways of connecting them to consumers; many tools for reconnection, however, depend on having sufficient numbers of successful local growers. Many of those I interviewed are taking the first small steps to work through this Catch 22. Massey University Sustainability Club's Farmers' Market was set up with the intention of providing students with access to locally grown, organic food, but the club has widened its net so that conventional growers can help provide the critical mass necessary to attract students to the organic growers.[36] Jones of RECAP speaks with similar enthusiasm about the development of a community market in Ashhurst. He is eager to see more produce at the market, but there are few local market growers available. Together with market organiser Helen King, Jones is therefore taking steps to organize a small group of home gardeners to plan together and plant surplus for a cooperative.[37] Another way of linking producers to consumers is illustrated through a community-supported agriculture (CSA) model adopted by Wairarapa Eco Farm. To give CSA farmers Frank van Steensel and Josje Neerincx a degree of financial certainty, customers subscribe as members and make regular payments. In return, they receive a harvest share in weekly food boxes. While CSAs have proven popular overseas (e.g. in North America and Europe), they are relatively uncommon in New Zealand, and the success of Wairarapa Eco Farm is important for demonstrating the viability of this type of farm-to-consumer connection. In all these cases there is a need to redevelop local growing and distribution step-by-step together, just as they have eroded and collapsed together.

As the links between the growing and eating of food are localized, reversing the trends associated with supermarket-based consumption, people become more deeply connected to the seasonality of food, and more aware of the entaglements involving soil, ecological functioning, and human health. This is underscored in a CSA, where members make an upfront choice to share risks with farmers. Wairarapa Eco Farm member Sita Venkateswar, who helps organize distribution to other Manawatu-based customers, told me of a year when everyone was looking forward to receiving a particular variety of summer stone fruit. Just before the fruit ripened, an unseasonal hailstorm decimated the crop. Even disappointments such as this, however, can help generate increased awareness of how our food ties us into connections with planetary processes that, though not beyond our influence and responsibility, are nonetheless beyond our control. Other ways of strengthening connections include member visits to the farm, newsletters, opportunities to work on the farm on a weekly basis in exchange for food, and learning the story of particular foods and of the particular soil in which it was grown. For Venkateswar, this is "coming back full circle" to the eating practices of her childhood in India, where everyone understood there was a connection between growing, harvesting, eating and health, with the right thing coming ripe at the right time.

With the emphasis on relocalization, even the smallest networks – such as garden sharing tables and neighbour-to-neighbour trade – are not only the beginning of efforts to become more food resilient; they are also an essential constituent of their ideal end. Such relations move at least some food out of the formal economy into the sphere of mutually reliant community relationship. "Food is free," says Madz BatachEl, the organizer of Manawatu Community Fruit Harvest (CFH), an all-volunteer group with a particular commitment to neighbour-to-neighbour distribution. On the surface, CFH's particular target is food waste, with volunteer fruit tree pickers harvesting fruit on a small-scale, domestic basis, generally from the homes of those who are unable to pick the fruit themselves. They then redistribute any surplus to those who can make use of it. Sometimes redistribution targets people who are socially disadvantaged, but not always; it's important to BatachEl, like several other network organizers, that some free food is disassociated from the stigma of food banks.[38]

BatachEl's putative role is to coordinate (1) volunteer pickers, (2) those with trees and food surplus, and (3) distribution points. Yet she has come to understand that she can make the biggest difference by teaching people skills for social (re-)connection, and her ideal is for CFH to no longer be needed. In a move that mirrors the deliberate decentralization of the original (Auckland-based) Community Fruit Harvesting, BatachEl puts some of her coordinating effort into supporting schools, libraries, and other organizations to set up sharing tables that require no further intermediary. She'd love to see food networks go "super-local," for example by giving people with trees mailbox flyers and encouraging or supporting them to ask their immediate neighbours to pick and share fruit. The goal of V. Witt's Trees for Homes project (distinct from Transition Feilding) is fully complementary: V. Witt donates fruit trees to private homes and plants them herself, using this as an opportunity for *in situ* education on tree care, companion planting, mulching, and other aspects of sustainable gardening. V. Witt's aim is to create a sustainable community orchard by relying on the ownership people will feel for the care of trees that are growing on their own property. She expects that, when the day comes that the trees produce too much fruit for one family, neighbours will share.

These organizers seek greater local self-reliance: Wouldn't it be great, asks Jones of RECAP, if Ashhurst and the Pohangina Valley were self-sufficient in food, water, and wine? But there is more to it. The hope that we can bring part of our growing and eating from the "pure" market into family and community interactions pervades many of these interviews. Venkateswar, for example, talks about CSA participation as a committed way of living one's values. There is a type of accountability that takes place when market connections, human relationships, and ecological feedback loops start overlapping more tightly in a local community. In the end, effective relocalization will require a wide range of strategies, from cooperatives, to community projects oriented towards sharing and exchange, to small enterprises.

Each of these approaches requires a different type of shift in consciousness and a different set of experiences. V. Witt explains. Transition Feilding's only fundraiser is their monthly attendance at a community market. Their market stalls serves as a point of outreach to a more diverse public, and it also generates some small funds through the sell of surplus seedlings, eggs or other produce. Transition Feilding also ensure they always have something free on hand, such as packets of seeds. They have chosen to call their garden a "sharing garden" to distinguish it from community garden models where only the people who work may harvest. People are always asking why Transition Feilding choose not to have an honesty box at their garden's sharing table: "We're giving people experience of different ways of exchanging." People get nervous at the idea of stuff being free, and V. Witt wants them to experience that nervousness and to think about why surplus is free while materials the group requires to keep growing aren't even for sale. Through such small steps, people are asked to interrogate their attitudes and how these attitudes influence social relations, with the idea that people will then rework their community and market connectivity in line with their change in consciousness.

RECONFIGURING COLLECTIVES AS AN ETHICS OF ACTION

> "I think the mere act of voicing or outlining a vision for the future can work to activate the process [of change], even if it takes some time (years). … One or two people can serve as triggers and then all sorts of change starts happening around them."[39]

When I asked if she thinks what Plant to Plate are doing contributes to larger social change, Duker expressed a rare moment of deflation, leaning back in her chair for the only time in the interview. "I sure hope it does." When I then asked her to describe her hope, her customary enthusiasm returned. She hopes that what Plant to Plate are doing will open the eyes of people who don't yet know the cooking and gardening skills they're teaching. "When working with young kids in bad situations," such as poverty, Duker wants them to receive "even just a few moments of connection with their environment," or "knowing that they *can* cook." Groups working with youth and young adults, such as Palmerston North Girls' High Enviro Group and Massey University Sustainability Club, are similarly creating a foundation for a life-long change of consciousness. "If you can get the message to people while they're still young – or at least when they have an open mind – they'll go out into the community and show other people."[40] There's a ripple effect when people are brought together in these initiatives: "There's a community of people all wanting to act."[41]

To Vanessa Witt, the grass roots is not only important: it is central. The only pathway to change, she emphasizes, is the individual working collectively. She recounts for me the metaphor of imaginal cells transforming a caterpillar to a butterfly, in spite of counteraction by the caterpillar's immune system. Maybe the whole society isn't transforming yet, concurs, Jill Faulkner, another Transition Feilding organizer, but things are now changing in multiple locations, and the time will come when multiple localized actions merge into a new society. V. Witt adds that we're always talking about climate change tipping points, but there are also social change tipping points. She tells me that Transition Feilding and the long-established Keep Feilding Beautiful (KFB) have started supporting one another, in spite of having distinctive aims ("veggies in gardens [for Transition Feilding], not flowers"). This year, KFB have, for the first time, included a sustainable and permaculture gardening category in their annual gardening competition. Similar roll-out impacts are part of Paewai's vision for the Parahaki Island māra kai. If the garden is, above all else, an authentic way to provide healthy food and environmental and community connection for hapū, then tourists can also come by, learn a bit of history, learn about traditional food plants such as kamo kamo, and increase their understanding of how to contribute to the garden revival in their own home.

For at least some in the network, the very smallness of community food initiatives is critical to their success, with the issue of globalization not only one of distance but also of scale. Venkateswar tells me that global problems mandate small-scale solutions of all types: time banks, savings pools, berm gardens, neighbourhood-owned small-scale electricity grids, and more: when small, such initiatives remain "transparent, accountable, and face-to-face"; whereas when they grow, "there is more chance for them to mutate into something else," something less likely to be at the service of human communities. "In the end the whole structure of global corporations hinge on consumption, so if the way you consume shifts, it has to start making a dent." The capitalist logic of consumption, Venkateswar continues, has made inroads into all aspects of our life, and it is now our task to see how this logic has been internalized in our consciousness, our lifestyle patterns, and our systems, so that we can more consciously choose from here on out. To Jones, there is no doubt that community initiatives will make a difference, a conviction that for him arises from years of action in union organizing. The trick, he tells me, is to create social points for whole groups of people to come together. In this way, only one or two people are needed to catalyse change all around them, drawing in others and forming new organizations. These in turn can create supportive environments for yet more people to raise their ideas and to develop first-hand experience of acting in community spaces.

Collectively, these initiatives are imbued with a sense of regeneration, with these on-the-ground initiatives are informed less by a spirit of resistance than by a spirit of co-creativity. Yes, these initiatives are enacted in a context of conscious awareness of global problems that require an urgent response. They are oriented towards comprehensive social change rather than the development of purely local pockets of alternative lifestyles. There is, in consequence, an element of resistance in this work, and it shows up in a range of daily experienced barriers and frustrations and the struggle to keep going when discouraged.

Overall, however, determination, love and joy are dominant characteristics of these organizers, with some positively radiant when they speak of their motivations. Their work is a way of reimagining and recreating our world for a better future. In this spirit, traditional ideological divides break down, and many of those I interviewed cooperate readily with other groups regardless of political affiliations or sector participation (i.e. government, business, education). In keeping with this emphasis on positive co-creation, I understand the activities of those I interview, as a "third way" – neither protest activism nor lifestyle activism – but something else where resistance is incidental to the co-creativity of communities that are simply coming together as humans, sunshine, rain, soil and seed to build the positive future for which participating human actors hope.

It's all about planting seeds. This metaphor has become entirely cliché, but its meaning is renewed through the actions of those I interviewed and through those of the many co-participants that create these local food initiatives alongside them. For these people, a central component of systems change is seeking out heritage seeds, teaching children how to plant and save seeds, including seed swaps as a celebratory aspect of volunteer thank you parties, and distributing free seed at information booths. At the grass roots, there is every reason to think these efforts, carefully adapted to each location – each very particular entanglement of humans and place – will keep on spreading, as surely as the grass itself, as surely as the grains grown for home-baked bread.

NOTES

1. Sita Venkateswar (Representative of Palmerston North Community Supported Agriculture [CSA] which serves as the local distribution group for Wairarapa Eco Farm), interview by author, December 2015.

2. Manahi Paewai (Elder and leader of Te Kāuru Eastern Manawatū Hapū Collective), interview by author, January 2016.

3. Vanessa Witt, Jill Faulkner, and Trevor Witt (Convenor, Secretary, and Treasurer, respectively, of Transition Feilding), interview by author, October 2015; Harvey Jones, (Chairperson of the Society for the Resilience and Engagement of the Community of Ashhurst and Pohangina Inc [RECAP]), interview by author, December 2015; Cindy Duker (Coordinator of Plant to Plate), interview by author, November 2015; Madz BatachEl (Coordinator of Manawatu Community Fruit Harvest), interview by author, October 2015; Venkateswar, interview.

4. Duker, interview.

5. Paul Jenkin (Past Chairperson of Manawatu Beekeepers' Club), interview by author, January 2016.

6. BatachEl, interview.

7. This phrase is taken from a permaculture principle inspiring to some of those interviewed. See David Holmgren, *Permaculture: Principles* and *Pathways Beyond Sustainability* (Hepburn, VIC: Holmgren Design Services, 2002).

8. Venkateswar, interview.

9. Forrest Chambers (community activist), personal communication, January 2009.

10. Environment Network Manawatu, *The Constitution of Environment Network Manawatu Inc* (Palmerston North: Environment Network Manawatu, 2001 [revised 2015]).

11. Nabhan, Gary Paul. *Enduring Seeds: Native American Agriculture and Wild Plant Conservation* (San Francisco: North Point Press, 1989).

12. Sandra Harding, "Rethinking Standpoint Epistemology: What is Strong Objectivity?" in L. Alcoff and E. Potter, eds., *Feminist Epistemologies* (New York/London: Routledge, 1993); Donna J. Haraway, *Modest_Witness@Second_Millennium. FemaleMan©_Meets_OncoMouse*[a]*: Feminism and Technoscience.* (New York and London: Routledge, 1997).

13. Solutions to group-level issues such as limited resources (funding, volunteers, access to land, etc.) and or inter- and intra-group conflicts are beyond the chapter's scope.

14. Witt, Faulkner and Witt, interview.

15. Bridgette Murphy and Eleanor Greer (Creative Director and member of Rangiwahia Environmental Arts Centre Trust [REACT], respectively), interview by author, November 2015.

16. Paewai, interview.
17. Kim Penny (Manager of SuperGrans Manawatu), interview by author, September 2015.
18. Palmerston North Girls' High (PNGHS) Enviro Group students (including Kaia Costanza-van den Belt and students who preferred to be unnamed), interview by author, March 2016.
19. Yvonne Marsh (Trustee of Green Hub) interview by author, September and October 2015; Jones, interview; Witt, Faulkner and Witt, interview.
20. Marsh, interview.
21. Witt, Faulkner, and Witt, interview.
22. Ibid.
23. Jenkin, interview.
24. Antonia McGhie (President of Massey University Sustainability Club) interview by author, November 2015.
25. Duker, interview.
26. Marsh, interview.
27. Witt, Faulkner, and Witt, interview.
28. Duker, interview; Penny, interview.
29. Jones, interview.
30. Ibid.
31. Penny, interview.
32. Ibid.
33. Ibid.
34. BatachEl, interview.
35. Charles Eisenstein, *Sacred Economics: Money, Gift and Society in the Age of Transition* (Berkeley: Evolver Editions, 2011).
36. McGhie, interview.
37. Jones, interview.
38. Witt, Faulkner, and Witt, interview; Jones, interview.
39. Jones, interview.
40. McGhie, interview.
41. Ibid.

BIBLIOGRAPHY

BatachEl, Madz. Coordinator of Manawatu Community Fruit Harvest. Interview by Author. October 2015.

Chambers, Forrest. Community activist. Personal Communication. January 2009.

Duker, Cindy. Coordinator of Plant to Plate. Interview by Author. November 2015.

Eisenstein, Charles. *Sacred Economics: Money, Gift and Society in the Age of Transition*. Berkeley: Evolver Editions, 2011.

Environment Network Manawatu. *The Constitution of Environment Network Manawatu Inc*. Palmerston North: Environment Network Manawatu, 2001 [revised 2015].

Haraway, Donna J. *Modest_Witness@Second_Millennium. FemaleMan©_Meets_OncoMouse[a]: Feminism and Technoscience.* New York and London: Routledge, 1997.

Harding, Sandra. "Rethinking Standpoint Epistemology: What is Strong Objectivity?" *Feminist Epistemologies.* Eds. L. Alcoff and E. Potter. New York and London: Routledge, 1993.

Holmgren, David. *Permaculture: Principles* and P*athways Beyond Sustainability.* Hepburn, VIC: Holmgren Design Services, 2002.

Jenkin, Paul. Past Chairperson of Manawatu Beekeepers' Club. Interview by Author. January 2016.

Jones, Harvey. Chairperson of the Society for the Resilience and Engagement of the Community of Ashhurst and Pohangina Inc (RECAP). Interview by Author. December 2015.

Marsh, Yvonne. Trustee of Green Hub. Interview by Author. September and October 2015.

McGhie, Antonia. President of Massey University Sustainability Club. Interview by Author. November 2015.

Murphy, Bridgette and Eleanor Greer. Creative Director and Member of Rangiwahia Environmental Arts Centre Trust (REACT), respectively. Interview by Author. November 2015.

Nabhan, Gary Paul. *Enduring Seeds: Native American Agriculture and Wild Plant Conservation.* San Francisco: North Point Press, 1989.

Paewai, Manahi. Elder and Leader of Te Kāuru Eastern Manawatū Hapū Collective. Interview by Author. January 2016.

Palmerston North Girls' High (PNGHS) Enviro Group Students (including Kaia Costanza-van den Belt and students who preferred to be unnamed). Interview by Author. March 2016.

Penny, Kim. Manager of SuperGrans Manawatu. Interview by Author. September 2015.

Venkateswar, Sita. Representative of Palmerston North Community Supported Agriculture. Interview by Author. December 2015.

Witt, Vanessa, Jill Faulkner, and Trevor Witt. Convenor, Secretary, and Treasurer, respectively, of Transition Feilding. Interview by Author. October 2015.

Chapter 8

In Different Voices

Engaging with Human–Nonhuman Entanglements

Sita Venkateswar

Engaging as an anthropologist within the current global, eco-political conjuncture requires a kind of vocal ambidexterity entailing multiple modes of expression in writing. In this chapter[1] I experiment with ways of thinking and writing about human–nonhuman entanglements in relation to questions of food security within the conjuncture designated as the Anthropocene.[2]

Two different modes of fieldwork are harnessed in this chapter to develop the narratives presented here. In the first, a mode of immersive engagement designated as "memorywork," is nourished by imagination to extract long buried quotidian details of lives lived in times past. I start with a storied approach that draws on retellings of family stories to highlight how people are always already amidst multiple species of others, whether acknowledged or not in quotidian lives. The effort here is to construe a body of conceptual work within anthropology and articulate a sense of long-standing familiarity with a conceptual corpus that has gathered momentum across the humanities and social sciences over the past decade. The chapter then applies a mode of research that I describe as "cyber fieldwork" to elicit an archive of resources documenting changing consumption practices in India which enable comprehension of the significance of plants/crops such as millets. Unravelling the agro-ecology of millets to understand the basis for its attribute of resilience requires close attention to the ways in which human farming practices are implicated within lifecycles of multiple species of others – plant, animal and microbes – entangled in webs of symbiotic relationships that exist beyond human interventions, yet vulnerable to the consequences of the Anthropocene on the felicitous conditions[3] for *all* life.

STAGED WRITING (1): "MEMORYWORK" AND STORIED ENCOUNTERS

Drawing together the skeins of my thoughts for this chapter, the ideas I develop here originate in my memories of my grandmother's stories from long, hot afternoons in Calcutta spent lying beside her in the shaded cool of her bedroom. The whirring ceiling fan picks up the occasional gust of warm breeze from the shuttered windows to settle on my increasingly heavy eyelids. As I listen to her reveries of a distant village in Kerala, her reminiscent voice casts a dream-like spell, sowing the seeds that have remained buried for decades to finally find fertile ground and germinate at this conjuncture in the Antipodes. I imagine the possibilities for what Eduardo Kohn[4] so evocatively describes as "an anthropology of life" to situate the human as inextricably entangled with all sorts of others.

I recall two stories in particular, both of which sent a thrill of something which I now draw from philosopher Alfred North Whitehead[5] to understand as "prehensions" or the intuitive apprehensions of possibilities that prefigures a sentience to come. The first is an incident from the pioneering journeys of my great-grandfather through the dense jungle of the Palakkad hills during the last quarter of the nineteenth century. A player in the futures market of that conjuncture, his work entailed identifying and marking jungle tracts rich in spices for auction. It was a dangerous venture through a wilderness teeming with predators. On one of these trips, his path through the jungle intersected with that of a feared predator, stretched across a rocky outcrop of the Western Ghats. His eyes locked with the amber, unblinking gaze of a leopard camouflaged by the dappled shadows cast by the thick vegetation surrounding them. My great-grandfather stood stock still, then bowing his head and drawing his palms together, he intoned: "Revered elder, if it pleases you, grant me permission to cross your path." The leopard's amber gaze burnished his face, then, in a fluid movement, the animal stretched, yawned and disappeared into the surrounding jungle. My great-grandfather went on to make a fortune trading in spices, but never forgot to give homage to the leopard that permitted him to grow old to tell his tale.

The second story emerges from the context of my grandmother's household responsibilities as a daughter, and the daily round of chores allocated to her. At the centre of the courtyard was the household well, which provided for the family's daily water needs, and it fell to my grandmother to draw the water for the family kitchen. One day, as my grandmother drew water from the well, she thought she heard a hissing sound. Her mother also noted the same susurration as she drew water for her morning ablutions. As the murmurs of apprehension among the women in the household grew louder and reached the ears of my great-grandfather, he took it upon himself to investigate the matter. Peering carefully into the crevices of the large well, he spotted a King Cobra hidden in the mossy gloom of the walls. Drawing his palms together and bowing his head low, he addressed the cobra. He said: "Revered elder, I live in this household with many children. Why have you come here to live among us? This is not a suitable home for you." He filled a cup of milk and left it by the well then ordered everyone indoors. The cobra uncoiled itself to slither sinuously away from the well and was never seen again; the well was emptied of water then left to replenish itself from the aquifer that fed it.

How do I read these stories of a time past signifying a different epistème? One option is to follow a mode familiar to me, to offer a worldly[6] reading that critiques my great-grandfather as an unwitting agent of capital, complicit with the colonial transformation of the Malabar landscape, enacting the vestiges of a soon to be replaced premodern rationality. Such a reading would trace the entwined movements of colonial capital and the socio-economic transfigurations that it left in its wake, displacing successive generations of my ancestors from the periphery to the centres of colonial modernity. It would discern continuing patterns extending through the postcolonial transmutations of late capitalism, bringing me to the shores of a different peripheral destination in the Pacific, where the colonial and the postcolonial are locked in embrace (or mortal combat depending on the point of view)! Unexpectedly, it is also where rich seams of entanglements imbue the landscape offering the promise of an "alter-modernity" in which, as anthropologist Eduardo Viveiros de Castro demonstrates: "Different points of views emerge from the ways in which different bodies constitute different modes of relating to, inhabiting and being enmeshed in, their environments."[7]

As I tap into the "black milk"[8] of memories, I find a home in the radical imaginary of contemporary anthropological thinking. Let us return to my grandmother's stories and the encounters with predators in Malabar. The similarities in my great-grandfather's mode of address to the two creatures are striking. Whether in the "wild" spaces of the jungle, or the "domestic" space of his house, in each instance, respect conjoint with an unwavering assumption regarding the possibilities for communication, mark the presence and the encounter with animals in those contact zones. Umberto Eco remarks on quite a different kind of prevalent sensibility in his essay from the late 1980s "How to Speak of Animals" and I quote: "To render animals worthy of rescue they are humanized, toyified. No one says they are entitled to survive even if, as a rule, they are savage and carnivorous. No, they are made respectable by becoming cuddly, comic, good-natured, benevolent, wise, and prudent."[9] My grandmother's account confronts the predators' ability to take human life head-on but the mutuality of humans and animals, their entitlement to survive, thrive and cohabit the spaces where both humans and animals range is never in any doubt. Her stories presage the "worlding" by anthropologist

Philippe Descola and others who "attempt to make sense of the multiple ways according to which humans describe the world and what they do in it."[10]

In the piece of writing that I stage above, I travel across many seas and many worlds to trace and retrace concrescence of prehensions that link the multispecies lifeworld of colonial Malabar to the twenty-first-century pluriverse[11] of recent scholarly discourses.[12] I have drawn on deeply embedded family stories to recover (and hence avert the possibility of forgetting), modes of dwelling in the world, through enacting a narrative alert and attuned to the entanglements that [re]configure the "anthropologies of life." My grandmother's stories can be perceived as parables to offer glimpses of times, places and quotidian practices that remember human and animal worlds as interconnected, hence serve to "naturalise" and mutually affirm each other's right to exist alongside one another in the world.

IN PARALLEL TRAIN

Alongside my efforts to identify a personal point of entry into multispecies thinking, a related trajectory of research articulated lines of enquiry pertaining to questions of livelihood and food security in the face of climate change and the ways these issues evolve in the contemporary context of India. These lines of enquiry demand close attention to context at the micro-level, triggering intimacy with the domain of the empirical accorded a different kind of voice as illustrated in the narrative that follows below.

STAGED WRITING (2): WRITING THE EMPIRICAL THROUGH MILLETS IN INDIA

Against the backdrop of climate change with its accompanying unpredictable and extreme weather patterns, questions of food resilience have assumed escalating importance across the world. In India, the crisis in rural areas with farmer suicides[13] highlights the precariousness of agricultural livelihoods. There is growing consensus among scholars and civil society groups that agricultural intensification along the same lines is no longer a sustainable option to combat hunger or the projected increase in population in the decades to come.

To explore the implications of these issues I embarked on a new research project at the end of 2011, addressing perceptions of climate change in India and the ongoing efforts to mitigate its impact. I travelled to Wayanad, in northern Kerala,[14] a region famed for its forests and fertility but also the site of increasing instances of farmer suicides over recent years. During this fieldwork, I met NGOs and Adivasi[15] farmers in Kerala and academics in Karnataka, which led to my induction into a cyber-network called "Revitalising Rainfed Agriculture," ably managed by Nemani Chandrasekhar, a programme officer with the Watershed Support Services Network based in Secunderabad.

"Cyber-fieldwork"

Over the past four years, Nemani has sent me digests six days a week. Organized by categories such as News Headlines, Food Security, Millets, Organic, Seeds, Fisheries, Livestock, Video, Publications, each digest is compiled of all the news and media reports pertaining to those categories from India, policy analysis as well as scholarly publications from India and elsewhere addressing those categories. Stored in a folder on my email, the daily digests have kept me abreast of the various efforts underway in relation to agriculture and related matters in India. They have provided me with an exhaustive overview of the kinds of thinking on the issues that require attention when addressing rural India or the rural-urban interface. In India, 60 percent of the total land devoted to agriculture is rainfed, that is, 85 million hectares spread out over 177 districts is entirely reliant on rainfall. These rainfed regions provide livelihoods to nearly 50 percent of the total rural workforce and sustain 60 percent of the cattle population in the country.

MATERIALIZING MILLET

My immersion in the enormous volume of information that I have amassed since January 2012 has enabled me to learn about millets, a hardy grain or nutri-cereal as it is referred to in some Indian media, or, more disparagingly, as coarse cereals in other media. It has nourished people and livestock in several regions of Asia and Africa over many millennia. You may want to ask why I have chosen to focus my attention on that grain? In line with my interest in climate change, millets are one of the most climate compliant crops available. Projections of climate change in India warn of rising temperatures, decreasing rainfall and scarcity of water. These conditions will have a major impact on the cultivation of rice and wheat, and in view of the extreme temperature sensitivity of wheat, there are concerns that it will even disappear. The standing water[16] required for rice cultivation demands considerable quantities of water, which increases the production of methane, one of the greenhouse gases.

Millets, on the other hand, are adapted to a wide range of ecological conditions, are all-season crops cultivated round the year requiring no irrigation and a fraction of the rainfall required for paddy or other cash crops like sugarcane or banana. They also survive on poor soils less than 15 cms deep; hence, they are an asset in the expanding dry, arid conditions anticipated for the future. Moreover, they do not require synthetic fertilizers and when grown using traditional methods, they attract no pests, neither during cultivation nor during storage. The presence of millet in the Sahel in Africa, one of the harshest environments in the world, testifies to their extraordinary capacity to thrive under adverse conditions. Most importantly, millets are not just crops but cropping systems because most millet fields are inherently biodiverse, 6–20 crops are planted in the same space and at the same time, ranging from oilseeds, pulses, including amaranth for fuelwood and fibre. Most millets have edible stalks which provide fodder for cattle. Clearly, millets provide multiple securities on a range of fronts: food, fodder, livelihood and ecological, thus rendering them the crop of agricultural security.

An additional basis for my interest in millet is the fact that it is primarily Adivasi groups across India who consistently subsist on millet and hence an engagement with millets is an extension of my ongoing and long-standing research interests on indigenous peoples and politics. Interestingly, prior to my virtual encounter with millet through Nemani's digests, it had already entered my kitchen in New Zealand and is one among a range of grains that I regularly incorporate into my diet. Purchased from organic food stores in New Zealand, and available as flour, flakes or wholegrain, it sits on a shelf along with other "exotic" ancient grains imported into New Zealand from remote regions of the world and available for consumption by those making a deliberate decision to expand their culinary horizon, as I have, or to maintain a gluten-free diet. However, it remained unknown to me during all the years that I lived in India, never crossing the threshold of my South Indian household and milieu in which I was raised, despite their high nutritional value. I am led me to ponder the caste and class implications of such an estrangement, or the prevailing cultural hegemony in the perception of rice and wheat as the primary staples of India, given the importance of millet across the Deccan Plateau, in Rajasthan, Gujarat, Maharashtra, Karnataka and parts of Tamil Nadu, apart from its preponderance in Adivasi diets.

I discovered that the umbrella term of millet is not any singular grain and its relatives, but "comprise of a group of grasses that grow in arid and semi-arid tropics and form a functional or agronomic group rather than a taxonomic one."[17] Hence, consumers of the product labelled "Millet," available in New Zealand organic shops, remain unaware of which particular millet they have purchased and therefore in what form it is best cooked and eaten. Homogenization and globalization are linked processes and the erasure of cultural knowledge goes unmarked in the proliferation of the global commodity chain.[18] Yet, despite their nutritional content, the cultivation of these grains have dropped by 50 percent over the past few decades as suggested in this figure, with the sharpest fall occurring between 1980 and 1990. A report in *The Hindu* dated April 21, 2013, quotes a senior agricultural officer who says:

> Major consumers of millets were people in the lower strata of society and their access to rice and wheat brought in a total transformation in food culture. Besides, conversion of agriculture land for real estate business, migration of poor farm labourers to urban area are also major factors in the decline of cultivation of millets.

The quote crisply amalgamates the range of factors brought to bear which have impact on the cultivation and consumption of millets in India. These factors also correlate with the reports of the increasing incidence of malnutrition in India and the changing subsistence practices that continue to contribute to the prevalence of malnutrition. Unsurprisingly, as a senior agricultural officer, he makes no mention of the Green Revolution and the relentless promotion of monocultures of rice and wheat for intensive farming in a few selected resource rich areas under irrigated conditions. National policies and the Public Distribution System, which have promoted the consumption of rice or wheat to the exclusion of other more nutrient rich staples like millet, despite regional preferences for ragi, jowar and bajra, (all region specific names for types of millets) have played no small part in this situation. Another line of research to pursue is to follow the patterns of displacement of tribal people in the name of various national or corporate projects and correlate them with the prevailing malnutrition trends.

A subsequent trip to Odisha contributed further insight into the situation. I learned that children in tribal areas are sent to residential schools where they are provided with a rice-based diet,[19] one that provides a fraction of the nutrients that they would typically derive from the millets they ate when living with their families. More troubling is the dietary acculturation that occurs, children sent to such schools refuse to eat anything but rice when they return to their families. Of all the millets, it is production of the minor millets or the small millets that have most dramatically dropped.

Just as there is the "Great Tradition" and the "Little Tradition" within anthropological thinking regarding India, there are the major millets such as ragi, bajra and jowar, consumed within large regional blocs in the south and west of India. Then there are the minor millets, primarily cultivated and consumed by Adivasi groups. They are not important in terms of the overall world food production, but they are strategic in terms of their nutritional contribution, and their role in local agro-ecosystems. What, then, are the implications to livelihood and well-being of that most marginalized of demographic groups in India, namely Adivasi who grow millets for subsistence rather than commercial production? Included within this demography are poor farmers especially in dryland India who are owners of very poor lands. Much of the cultivable fallow and low fertility farms are handed to them through the process of land reforms. The only crops that sustain agriculture and food security on these lands are millets.

A flurry of national and regional interventions in India have drawn my attention in recent months in relation to millets, predominantly from southern India, marking differences in approaches to food security between the north and south of the country. They also manifest a deviation to the existing rice and wheat-based hegemony to issues of food security and nutrition in India since the Green Revolution. However, the deeper malaise remains unaddressed in the series of initiatives highlighted earlier. Who are the targets and beneficiaries of such interventions? Do they address the consumers of the minor millets whose sustenance derives primarily from that grain? An examination of the literature suggests otherwise and confirms the consumption practices of the urban middle class as targets for these initiatives.

There is a degree of irony in the enculturation to more traditional ingredients and eating habits directed towards the more entitled sectors of the Indian population, juxtaposed against the ongoing erasure of other traditional practices of those least able to withstand those consequences, such as Adivasi contexts across the country discussed earlier. However, if millets are included in the Public Distribution System, as suggested by news reports,[20] with the Food Security Ordinance and the possibility that farmers are provided with incentives or a viable purchase price for that crop by the Central or State governments, then that is indeed a significant shift in policy in India. The restoration of the importance of millet on the national horizon has the potential to open up a new path to thinking about food production in India that combines questions of resilience with food security. It is crucial to avoid the temptation to simply substitute monocultures[21] of millet for rice or wheat in the agricultural sector and then continue as usual by applying the existing model of intensive, chemical-input-driven agricultural production. The cultivation of millet offers the opportunity to rethink how we grow our food and to what end.

FUTURE POSSIBILITIES?

At this juncture, in a narrative that has recounted the entanglements of human-animal-plant (in its domesticated avatar as "crop"), some interesting directions present themselves when shifting our gaze to consider how questions of resilience might be extended to simultaneously encompass the well-being of humans and other/more than human. In a provocatively titled chapter

"We have never been human" from her book "When species meet," Donna Haraway writes eloquently of the ubiquity of microbial life form. She says:

> I love the fact that human genomes can be found in only about 10 percent of all the cells that occupy the mundane space I call my body; the other 90 percent of the cells are filled with the genomes of bacteria, fungi, protists, and such, some of which play in a symphony necessary to my being alive at all, and some of which are hitching a ride and doing the rest of me, of us, no harm. I am vastly outnumbered by my tiny companions; better put, I become an adult human being in company with these tiny messmates. To be one is always to become with many.[22]

Building on Haraway's work among others, sociologist Myra Hird[23] advances a co-evolutionary approach to the micro-cosmos. Hird develops a micro-ontology framed by processes of symbiosis and symbiogenesis as elaborated by biologists Lynn Margulis[24] and Dorian Sagan.[25] Privileging symbiosis and symbiogenesis as processes that generate new organs, tissues and species, Margulis and Sagan[26] offer an entirely different perspective to the evolution of organisms. Shifting the motor of evolution from gradual, random mutations toward the formation of individuals, what we have instead are temporal and spatial assemblages, formed and remade through processes of contagion and filiation. Such multilateral meshworks are only apparent during outbreaks of epidemics in recent times, such as bird and swine flu or the ebola virus, but they broker all life, as we know it. Our only quotidian embodied awareness is when the immune system sends signals through the states of ease, un-ease or dis-ease, as we inhabit and dwell in the spaces and places of our lives with several species of others, both visible and invisible.[27]

In reflecting on what renders millet so significant at this conjuncture, it is evident that its hardy attribute, namely its resilience, distinguishes it from other crop staples and renders it climate compliant. Such an attribute of resilience derives from the symbiotic relationships that extend between millets and other-than-humans in the food web, enabling it to remain a hardy crop.[28] Another way of articulating this feature is to say that the livelihoods of humans in relation to millets hinge on the symbiotic relationships that sustain the livelihoods of both humans and nonhumans. The cultivation of millets has always been part of an integrated cropping system together with companion plants with their own complex webs of symbiotic relationships across a whole array of other life forms, both visible and invisible.

Such complex webs of symbiotic relationships were radically disrupted or destroyed by the interventions related to the Green Revolution. The systems of aggressive and intensive agricultural production are to the detriment of other life forms, stripping the soil of its vitality. The drastic impact on the environment of such a mode of production and to the relationships that sustain life have become clearly manifest over the past decades and are a major factor in the current crisis in the agricultural sector in India with the consequences to rural livelihoods.[29] In the constitution of humans and nonhumans and the webs that bind them, a multispecies approach to conceptualizing food production and cultivation is exemplified by millets.[30] What I intend to have accomplished by this stage is to interrogate how the human is typically constituted as bounded, autonomous entity, positioned in exceptional splendour in relation to other life forms. As queer theory estranged us from heteronormativity, contemporary thinking extending across anthropology, sociology, biology, philosophy and science studies both queries and queers the Anthropos, to foreground the lively "cosmo-pluriverse" in which we exist through intimate entanglements amidst others of all sorts. As historian Geoffery Lloyd points out in a recent paper, "What is at stake here is, among other things, how we should live. Ontology here, as so often elsewhere, implicates morality."[31]

CONCLUSION

Such multispecies, symbiotic commingling through congeries of relationships that I have sketched above, prefigures an ethics of engagement that is otherwise and truly as Levinas[32] intended, since the radical alterity of these teeming others precludes assimilation to the human and is necessarily non-anthropocentric. What we have instead is the conditions of possibility for "an anthropology of life," enabling us to reconceptualize the world we inhabit with a multitude of others and decipher our respons-a-bilities attuned to our multiple, irrevocably, interdependent worlds. Along with Haraway, Hird and also endorsed by recent scientific research on the significance of the "microbiome" to human health and well-being, I have que[e]ried the material constitution of "human" as separate from other life forms, and conventionally construed as a part of "nature." If the very condition of human life is always already inclusive of and hinges on the alterity of multiple species, how might such an awareness of our multispecies affiliations reconfigure our moral horizon and practice to render us *other*wise, and therefore capable of reckoning with the emergent lineaments of the "nonhuman human"[33] demanded from the twenty-first-century Anthropocene? As we imaginatively dwell among all the others who inhabit us, we are offered an embodied worlding to our understanding of the Maori concept of turangawaewae,[34] enabling us to stand in company with others to become one among many across species.

The interweaving narratives in this chapter set out parallel trains of conceptual and empirical lines of enquiry recounted in differing voices and modes of writing, brought together as an amalgamated narrative and as a combined effort to speak the human-animal-plant-microbe nexus in different voices and registers. I shift from human entanglements within the realm of majestic predators to those that exist invisibly, yet are omnipresent within the interstices of all lifeforms. Such a passage is mediated by my engagement with plant food staples and the crucial questions of *how* food is cultivated and to what end. Both the narratives staged above converge to foreground the significance of expanding our horizon of awareness beyond the human to the macro- and micro-cosmos of life forms that determine *all* life as we know it.

NOTES

1. There is a possibility for a double reading of the title to this chapter, that is "indifferent" voices alongside the intended "in different." While the intended "in different" literally alludes to the transitions in voice coursing through the narrative, the possibility of the former offers an intriguing subtext that connects with the next note, regarding the alternate ways the Anthropocene can be construed including the rejection of those immanent possibilities.

2. Hannah Gibson and Sita Venkateswar, "Anthropological Engagement with the Anthropocene: A Critical Review," *Environment and Society: Advances in Research* 6 (2015): 19. The reference to the Anthropocene alludes to the currently contested geological time scale that marks an era when human induced changes determine the life course of the entire planet. Gibson and Venkateswar draw attention to the ways in which the concept of the Anthropocene "reveals its Janus-faced propensities, demanding attention as to how the concept is applied within local, regional, and global politics, yet, simultaneously includes an invitation to think otherwise." See also Donna Haraway et al., "Anthropologists Are Talking – About the Anthropocene," *Ethnos* 81, no. 3 (2016): 535–64.

3. Bruno Latour, "Summary of the AiME project – An Inquiry into Modes of Existence," *Bruno Latour.fr*, October 24, 2011, http://www.bruno-latour.fr/node/328.

4. Eduardo Kohn, *How Forests Think: Toward an Anthropology Beyond the Human* (Berkeley: University of California Press, 2013).

5. Alfred North Whitehead, *Process and Reality: An Essay in Cosmology* (New York: The Free Press, 1978).

6. See Edward Said, *The World, the Text and the Critic* (Cambridge, MA: Harvard University Press, 1983) for an elaboration of "worldly" critique.

7. Ghassan Hage, "Critical Anthropological Thought and the Radical Political Imaginary Today," *Critique of Anthropology* 32 (2012): 298.

8. Paul Celan's potent imagery from his poem *Death Fugue* leaves its lyrical trail in Karen Bolender, "R.A.W. Assmilk Soap," in *The Multispecies Salon*, ed. Eben Kirksey (North Carolina: Duke University Press, 2014).

9. Umberto Eco, "How to Speak of Animals," in *How to Travel with a Salmon and Other Essays* (New York: Harcourt Brace & Company, 1994), 214–15.

10. Philippe Descola, "Cognition, Perception and Worlding," *Interdisciplinary Science Reviews* 235, no. 3–4 (2010): 335.

11. Here I draw on both Isabelle Stengers' and Bruno Latour's extensive writing for this coinage.

12. As argued by indigenous spokespersons responding to the recent emergence of multispecies approaches, such prehensions are an integral component of indigenous lifeworlds of the past and present and imbue many landscapes across the world.

13. A. R. Vasavi, "Suicides and the Making of India's Agrarian Distress," *South African Review of Sociology* 40, no. 1 (2009): 124–38; A. R. Vasavi, "The Absent Agriculturalist," *Georgetown Journal of International Affairs* (Winter–Spring 2013): 133–41; A. R. Vasavi, "Debt and Its Social Entrapment," *Women's Studies Quarterly* 42, no. 1&2 (2014): 23–37.

14. These are adjoining forests to the ones traversed by great-grandfather during the late nineteenth century. Hence, my insertion into these sites is also derived from what I perceived as long-standing relationships with that context and an effort to comprehend the series of crises that beset a region long known for sustaining the people who lived there.

15. In the Indian context, the term *Adivasi*, or "original inhabitant," is a Hindi word used by the general population, often as a pejorative, when describing aboriginal groups. It is meant to be synonymous with their "primitive" condition and marginal status. But those referred to as *Adivasi* are also included within the list of *Scheduled Tribes*, which in turn confers certain rights which is part of the affirmative action apparatus written into the Indian constitution. However, each group has its own name derived from their own language families to refer to themselves. *Onge, Munda, Nayaka* are some of the names of groups scattered across India and the Andaman Islands, each referent encompassing very specific linguistic, cultural and geographic attributes. Within the current political climate in India the "original" connotation of the *Adivasi* is ignored or contested since all Hindu Indians also assert their "originary" status based on the scriptures.

16. It is important to realize just how ancient and deeply embedded modes of rice cultivation are within diverse landscapes across south and south-east Asia. The ponding of water is part of a complex agroecology that requires a holistic and systems approach, or in more contemporary terms, a close attention to the varied assemblage that constitutes rice farming.

17. Aamchi Sheti, "Crops-Bajari," *Aamchi Sheti*, May 22, 2013, http://www.aamchisheti.vdfindia.org/Crops.html.

18. The linguistic diversity and the widespread use of millets across India is discussed in Ashwini Raghupathy, "Millets: That Wonder Grain!" *The Alternative.in*, August 25, 2012, http://thealternative.in/environment/millets-.

19. It is important to note that traditionally, parboiled rice and a range of rice species cultivated during different seasons provided an array of nutrients. When highly processed and polished rice from hybrid varieties dominated the market and consumption choices, following on the heels of the Green Revolution, malnutrition trends also became increasingly visible.

20. Deccan Herald, "Government Ready to Include Ragi, Jowar under PDS," *The Deccan Herald*, July 10, 2013, http://www.deccanherald.com/content/344080/government-ready-include-ragi-jowar.html.

21. It is instructive to recall that prior to the shifts in agriculture brought about by the Green Agriculture, the cultivation of rice and wheat in India was part of intercropping systems derived from a diversity of seed types. See Debal Deb, "Vrihi: An Effort to Conserve Folk Rice Varieties," *SlideShare*, April 21, 2013, http://www.slideshare.net/DebalDeb/vrihi for more details of Deb's work to retrieve and restore the diversity of rice seeds available across different region of India. See also Debal Deb and K. C. Malhotra, "Conservation Ethos in Local Traditions: The West Bengal Heritage," *Society & Natural Resources* 14:8 (2001): 711–24.

22. Donna Haraway, *When Species Meet* (Minneapolis: Minnesota University Press, 2007), 3.

23. Myra Hird, "Coevolution, Symbiosis and Sociology," *Ecological Economics* 69 (2010): 737–42; Myra Hird, "Meeting with the Microcosmos," *Environment and Planning D: Society and Space* 28 (2010): 36–39.

24. Lynn Margulis, *Symbiosis in Cell Evolution* (New York: W.H. Freeman & Co. Ltd, 1981).

25. Dorian Sagan, "The Human is More Than Human: Interspecies Communities and the New 'Facts of Life'," *Cultural Anthropology*, April 24, 2011, http://www.culanth.org/fieldsights/228-the-human-is-more-than-human-interspecies-communities-and-the-new-facts-of-life.

26. Lynn Margulis and Dorian Sagan, *Acquiring Genomes: A Theory of the Origin of Species* (New York: Basic Books, 2002).

27. *Cascade Experiments* is a poem by Alice Fulton, the fractal poet who alludes, to the unseen, invisible, unspoken yet teeming with a complexity requiring considerable effort to comprehend, and like the Anthropocene, is an invitation to think otherwise. "The poetics I've termed 'fractal' is concerned with the discarded details – overlooked aspects of structure, and more importantly, the dark matter of content: knowledges that remain unspoken because to attend to them would be inconvenient, if not dangerous. In such a way the tenets of complexity theory can suggest ethical as well as aesthetic dimensions." http://www.alicefulton.com/books/isr2005.html. Alice Fulton's work features in Karen Barad's writing, a feminist physicist who writes on the philosophy of science and who is included among the neo-materialists, who offer a new metaphysics that "traverses and thereby rewrites *thinking as a whole*, leaving nothing unchanged, redirecting every possible idea according to its new sense of orientation." Rick Dolphijn and Iris van der Tuin, *New Materialism: Interviews and Cartographies* (Ann Arbor, MI: Open Humanities Press, University of Michigan), 13.

28. For an example of a rice farm food web comprising of above ground and visible life forms see Debal Deb ibid., slide #9. My own unfolding millet(s) research involves working with agroecologists and agricultural scientists to assemble and map a similar web of life with a point of difference; it will include the relevant soil micro-organisms, particularly mycorrhizal fungi within such a map. See also Haraway et al. ibid., for more discussion of the critical significance of collaborative research teams extending across the humanities, social sciences and sciences at this juncture. Anna Tsing's

ongoing research exemplifies such a stance. See particularly her research on the matsutake mushroom *The Mushroom at the End of the World: On the Posssibility of Life in Capitalist Ruins* (Princeton: Princeton University Press, 2015) and the research she leads with the AURA group (Aarhus University Research on the Anthropocene).

29. Jennifer J. Onorati, "How Local Knowledge Can Restore the World's Food Supply: An Ethnoecological and Political Ecological Critique of Industrial Agriculture," December 10, 2012 (draft paper, Academia.edu).

30. Ibid.; Gibson and Venkateswar, 12–13 for more discussion on the "posthuman."

31. E. R. Geoffery Lloyd, "Humanity Between Gods and Beasts? Ontologies in Question," *Journal of the Royal Anthropological Institute (N.S.)* 17 (2011): 838.

32. Emmanuel Levinas, *Otherwise Than Being, Or, Beyond Essence* (The Hague: Nijhoff, 1981) invoked the otherwise "to refer to that which is ungraspable, unobtainable, and other than what *is*. By positing the 'other' as something out of our reach, Levinas's work challenges the Enlightenment idea that all things can be known in the world." Ibid.; Gibson and Venkateswar, 20. See also Elizabeth A. Povinelli, "Geontologies of the Otherwise," *Cultural Anthropology Online*, January 13, 2014, http://culanth.org/fieldsights/465-geontologies-of-the-otherwise; Elizabeth A. Povinelli, "After the Last Man: Images and Ethics of Becoming Otherwise," *E-Flux Journal*, no. 35, May 2012, http://www.e-flux.com/journal/after-the-last-man-images-and-ethics-of-becoming-otherwise/; Elizabeth A. Povinelli, "Routes/Worlds," *E-Flux Journal*, September 27, 2011, http://www.e-flux.com/journal/routesworlds/.

33. Giorgio Agamben, *The Open: Man and Animal* (Stanford: Stanford University Press, 2003).

34. *Turangawaewae* is a profoundly emotive Māori concept that connotes the place that one can claim as one's own to stand and act in the world.

BIBLIOGRAPHY

Agamben, Giorgio. *The Open: Man and Animal*. Stanford: Stanford University Press, 2003.

Bolender, Karen. "R.A.W. Assmilk Soap." *The Multispecies Salon*. Ed. Eben Kirksey. Durham: Duke University Press, 2014.

Deb, Debal. "Vrihi: An Effort to Conserve Folk Rice Varieties." *SlideShare*, April 21, 2013, http://www.slideshare.net/DebalDeb/vrihi.

Deb, Debal and K. C. Malhotra. "Conservation Ethos in Local Traditions: The West Bengal Heritage." *Society & Natural Resources* 14:8 (2001): 711–24.

Descola, Philippe. "Cognition, Perception and Worlding." *Interdisciplinary Science Reviews* 235:3–4 (2010): 334–40.

Dolphijn, Rick and Iris van der Tuin, *New Materialism: Interviews and Cartographies*. Ann Arbor, MI: Open Humanities Press, 2012.

Eco, Umberto. "How to Speak of Animals." *How to Travel with a Salmon and Other Essays*. New York: Harcourt Brace & Company, 1994.

Gibson, Hannah and Sita Venkateswar. "Anthropological Engagement with the Anthropocene: A Critical Review." *Environment and Society: Advances in Research* 6 (2015): 5–27.

Hage, Ghassan. "Critical Anthropological Thought and the Radical Political Imaginary Today." *Critique of Anthropology* 32 (2012): 285–308.

Haraway, Donna. *When Species Meet*. Minneapolis: Minnesota University Press, 2007.

Haraway Donna et al. "Anthropologists Are Talking – About the Anthropocene." *Ethnos* 81:3 (2016): 535–64.

Herald, Deccan. "Government Ready to Include Ragi, Jowar under PDS." *The Deccan Herald*, July 10, 2013, http://www.deccanherald.com/content/344080/government-ready-include-ragi-jowar.html.

Hird, Myra. "Coevolution, Symbiosis and Sociology." *Ecological Economics* 69 (2010): 737–42.

———. "Meeting with the Microcosmos." *Environment and Planning D: Society and Space* 28 (2010): 36–39.

Kohn, Eduardo. *How Forests Think: Toward an Anthropology Beyond the Human*. Berkeley: University of California Press, 2013.

Latour, Bruno. "Summary of the AiME Project – An Inquiry into Modes of Existence." *Bruno Latour.fr*, October 24, 2011, http://www.bruno-latour.fr/node/328.

Levinas, Emmanuel. *Otherwise Than Being, Or, Beyond Essence*. The Hague: Nijhoff, 1981.

Lloyd, E. R. Geoffery. "Humanity Between Gods and Beasts? Ontologies in Question." *Journal of the Royal Anthropological Institute (N.S.)* 17 (2011): 838.

Margulis, Lynn. *Symbiosis in Cell Evolution*. New York: W.H. Freeman & Co. Ltd, 1981.

Margulis, Lynn and Dorian Sagan. *Acquiring Genomes: A Theory of the Origin of Species*. New York: Basic Books, 2002.

Onorati, Jennifer J. "How Local Knowledge Can Restore the World's Food Supply: An Ethnoecological and Political Ecological Critique of Industrial Agriculture." December 10, 2012 (draft paper, Academia.edu).

Povinelli, Elizabeth A. "Routes/Worlds." *E-Flux Journal*, no. 27, September 2011, http://www.e-flux.com/journal/routesworlds/.

———. "After the Last Man: Images and Ethics of Becoming Otherwise." *E-Flux Journal*, no. 35, May 2012, http://www.e-flux.com/journal/after-the-last-man-images-and-ethics-of-becoming-otherwise/.

———. "Geontologies of the Otherwise." *Cultural Anthropology Online*, January 13, 2014, http://culanth.org/fieldsights/465-geontologies-of-the-otherwise.

Raghupathy, Ashwini. "Millets: That Wonder Grain!" *The Alternative.in*, August 25, 2012, http://thealternative.in/environment/millets-.

Sagan, Dorian. "The Human is More Than Human: Interspecies Communities and the New 'Facts of Life'." *Cultural Anthropology*, April 24, 2011, http://www.culanth.org/fieldsights/228-the-human-is-more-than-human-interspecies-communities-and-the-new-facts-of-life.

Said, Edward. *The World, the Text and the Critic*. Cambridge, MA: Harvard University Press, 1983.

Sheti, Aamchi. "Crops-Bajari." *Aamchi Sheti*, May 22, 2013, http://www.aamchisheti.vdfindia.org/Crops.html.

Tsing, Anna. *The Mushroom at the End of the World: On the Posssibility of Life in Capitalist Ruins*. Princeton: Princeton University Press, 2015.

Vasavi, A. R. "Suicides and the Making of India's Agrarian Distress." *South African Review of Sociology* 40:1 (2009): 124–38.

———. "The Absent Agriculturalist." *Georgetown Journal of International Affairs* (Winter–Spring 2013): 133–41.

———. "Debt and Its Social Entrapment." *Women's Studies Quarterly* 42:1–2 (2014): 23–37.

Whitehead, Alfred North. *Process and Reality: An Essay in Cosmology*. New York: The Free Press, 1978.

Part III

EPISTEMOLOGY, AESTHETICS AND MEDIATION

Chapter 9

Photographic Reflections on Landscape Change in Regional Australia

Christopher Orchard and James Holcombe

This chapter takes shape as a kind of unfolding and entangled dialectic, not between people, but between disciplinary perspectives on historical and contemporary nature/culture binaries. It addresses shifting Australian perceptions of land over time through an interdisciplinary approach to critically engaging the spheres of ecology, psychology and photography through a chronological assessment of Australia's changing national identity. These movements are loosely grouped in to modes of colonization, the birth of nationhood, early environmental-photography and political activism, and finally a post-humanist post-materiality conceived in terms of photography for environmental mutualism. In tracing these disciplines, the narrative shifts towards what might be considered a new artistic movement (or moment) in the definition of a dialogic of creative practice for the Anthropocene. It does so to equally trace the emergent discourse of ecopsychology, a branch of environmental psychology and environmental philosophy principally concerned with addressing human–nonhuman disconnects and reconnecting the mind meaningfully with the world about it. Equally this chapter searches for ways that addressing this disconnect might open new sources for meaning making and psychic equanimity towards intergenerational equality of opportunity. It is asking for nothing less than to change how we view our place in the world, and how we might come to record that view.

REJECTING LANDSCAPE TRADITIONS AND EMBRACING AN ENVIRONMENTAL MUTUALISM

The lingering western traditions of colonization and nature/culture binaries reverberate in contemporary landscape photography, like ghosts of an agrarian mentality, through the ongoing collecting and commodifying of landscape, culture and experience. Through photography, these seemingly habitual behaviours provide many in Australian society continued excuse to practically and rhetorically avert their gaze from the realities of ecological disaster and biodiversity loss; particularly in regard to that of the threats faced from anthropogenic climactic change. This aversion is logically dishonest, psychologically problematic and principally unethical, and results in an ongoing exploitation of the documented world by commercial and many fine-arts photographic practitioners.

Liz Wells identifies the photographic image in *Photography a Critical Introduction* as "a coded artefact, capable of being read as cultural, psychoanalytic and ideological signs."[1] In order to generate a contemporary photographic and post-photographic practice, which rejects the anthropocentrism and responds to the interventionism of Australia's past, it is necessary to reference a new environmentally and ethically focused aesthetic. This would not objectify nature as a cultural construction at the hand of capitalist exploitation, but would rather embed the photographer within the natural rhythms and cycles of life, to respond to them, from within them. The aim is to generate and articulate a photographic approach that is ultimately aware of its cultural, psychological and ideological pedigree, which is both explicit and implicit in its practice. Its subsequent wide-ranging impact can potentially heal our current nature/culture binary, by reinstating a practice-led methodology through a being within the world.

Any new aesthetic system for the Anthropocene has an obligation to be ethically robust, socially and culturally responsible, and community embedded, while simultaneously contributing to a meaningful long-term psychological health and well-being towards intergenerational equality of opportunity. It is suggested that an appropriate nomenclature for the principles of this practice is photography for environmental mutualism. This eco-aesthetic comes with a primary acknowledgement that photographers are in no way separate from the living systems they aim to record, and indeed they are accountable to this complex symbiotic organization of relationships. This knowledge, when born-out through artefact generation, should form part of a culturally subversive duty of the photographic art form to bring ecological crises, on various scales, to a broad social awareness.

This requires a repositioning of the photographer away from conventions of photographic colonizer and intervener and towards the role of ethical steward. Such a model requires the traditional landscape photographer to remove themselves from both the historic religious glorification and the institutionalized anthropocentrism that continues to plague the landscape genre, particularly, in the case of the latter, the commercialization of land. Challenging the historic process of commercialization also requires modes to address the simplification of earth systems, the psychology without ecology, and the domestication of plants, animals and other nonhuman actors. It asks photographers to reposition themselves within the broadest terms of the natural, to focus on being and working within nature: to recognize that to infantilize the world about us is to emotionally and psychologically cripple ourselves. In this new mode of working, the commercial or fine art photographer is a living animate body, which is meaningfully connected to its environment (as part of the ordinary state of being), and becomes a conduit of expression for a broad environmental and ecological web. This mode is beyond traditional landscape photography, such as the picturing of a view or the aestheticizing of land, where the photographer emerges as contemporary documentarian. Equally, this is photography beyond gallant frontiersmen placing themselves as heroes at a territory's centre, acting as architects or obstacles of that territory, and becoming solitary narrators of landscape histories and their potential futures. Rather, this practice calls for a more altruist approach from an active experiencer functioning in a wider system that also lives, conveying that system's voice on its own terms. To speak in allegory, it might ask how a rock, river, desert or tree would prefer to be illustrated and valued, and that their voice through us might tell us and communicate to us of our own psychological well-being; despite psychologies prevailing reality principle considering no greater madness "than to believe that beast and plant, mountain and river have a point-of-view."[2]

This photography for environmental mutualism draws no boundaries between photography, ecology and psychology. Instead, more like an overlapping Venn diagram, the three principal disciplinary spheres overlap in increasing occurrence at the centre. This generates an interdisciplinary mode of making that breaks down the old boundaries of self/otherness and other modern anthropocentric psychological constructs. The argument here is for a kind of consilience of thinking, in which traditional discipline boundaries converge with stronger claims and resolutions for action. The practitioner is asked to act always in service to the betterment and sustainment of the communities or bioregions in which they live and function. The viewer is likewise asked to undertake the burden of accepting his or her own ecological, cultural and economic footprint (including intergenerational liability), and to own the experience as an authentic model towards increased knowledge, which can be shared as active political action. This is a photography that is bound to advance society in making the unconscious visible through what ecopsychologists might call a kind of latent-indigenous knowledge of our planetary relations, a relationship that ecopsychologists, like Paul Shepard and Ralph Metzner, are has been systemically repressed over generations since the invention of agriculture.[3] In recognizing the principle dishonesty in separating the human from the world about us, we can steer society back toward effective and operational political leadership for psychic equanimity and intergenerational equity. It acknowledges that our relationship to experience (sometimes place) is important to our mental well-being, and asks photographers, as ideological, psychoanalytical and cultural documentarians, to be honest and informed in that relationship.

CONSILIENCE THINKING

Within higher education, terms like "interdisciplinarity," "transdiciplinarity" and "multidisciplinarity" generally remain obfuscatory weasel-words, but they do have merit when applied with the right intentions. They can represent meeting points of modes of thought that can be fundamental places for the revitalizing of traditional disciplines, and this is ultimately consilience. Psychology on its own, for example, has traditionally disconnected the human thinker from the ecological systems that are external to human thought, leaving little room for interconnectedness. Photography, by holding up a mirror on our society and the environment, has the capacity to allow us to critically reflect on the folly of such a disconnection. When utilized this way, the photograph houses evidence of a disruption in the usually integrated functions of consciousness, memory, identity and environmental perception. Traditional photographic landscape practices tend to deny these attributes though, by covering the mirror and subsequently contributing to dissociation and continuing the myth of the human disconnected from its environmental responsibility.

In *Spectral Evidence*, Ulrich Baer is clearly addressing these ideas of denial and dissociation by referring to terms such as *non-seeing* and *theft of experience*. Such non-seeing becomes a social document that acts as a deformation of memory and temporality: reacting in outbursts of unresolved and disconnected experience. This experience is a mixture of familiarity and anxiety, nostalgia and expectance, suggestion and dissociation and, as such, the landscape photograph is a documentation of an ecological trauma. Trauma, like the photograph, is a disorder of memory and time; its experience is deferred and arrested in the unconscious, awaiting its movement or development into the conscious. The photographic landscape image, as a ubiquitous cultural vessel, has become a receptacle for our collective nostalgia for the past, our dissociation from the present, and our anxieties for the future, and at its core it speaks of a cultural aversion to acknowledging a shared ecological trauma.

MODES OF COLONIZATION: PHOTOGRAPHY, ECOLOGY AND PSYCHOLOGY

It is possible to analyse the photography, ecology and psychology of colonization from a regional Australian perspective – which acts as a microcosm for global psychopathological and environmental concern – in order to find a way forward from the separation of humanity and nature in photography of the environment. Much of early Australian photography from the mid-nineteenth and early-twentieth centuries was undertaken as a process of continued pioneer documentation and exploration. The aesthetic modes of the sublime and picturesque imported with early European settlers predominated photographs of the Australian landscape. These modes served to provide either a depiction of the unknown outback as a rousing wilderness fit for future English exploration, or a means of normalizing the land as hospitable in order to increase European agricultural and pastoralist interest. Many early pioneers were of convict pedigree; the term "mate" is derived from shipmate. Convicts transported aboard English vessels to Australia were barred from addressing each other by their given names, and as such, mate became common parlance in the colonies to refer to any other convict. This term, through repetition and necessity, stuck in the Australian vernacular and became suggestive of a social class, one of ex-convict Australians attempting to escape their criminal background through finding new wealth and prosperity in a new land, and raising their social caste.

Anne Marie Willis notes in *Picturing Australia: A History of Photography* that early Australian landscape photography depicted the Australian bush as picturesque and pleasing to behold. Ultimately this glamorized the inhospitable by displaying a new kind of rugged beauty captured as commodity. In *Photography and Landscape*, Rodney Giblett adds that this kind of photography made Australia "ripe for settlement by portraying it as a pleasing prospect for immigration and agriculture."[4] This agricultural expansionism principally saw "the planet as a thing rather than a thou, a product instead of an organism, to be possessed rather than encountered as a presence."[5]

Images of an expansive and available Australia can be seen is the works of pioneer settlers Morton Allport, the Anson Brothers, Charles Kerry & Co, and later in the nineteenth century in the work of Stephen Spurling III and John Watt Beattie. Photographs of prime agricultural land by Townsend Duryea, George William Townsend, Cuthbert Fetherstonhaugh and numerous photographers on assignment from the Sydney Mail pictured the Riverina region of New South Wales as prime land for agricultural expansion. So lingering is this photographic legacy of expansionism that currently inscribed in brass atop Galore Hill in the Lockhart Shire, one of the Riverina region's few elevated vantage points, resides the early white-explorer Henry Osborne's epigraph: "There's land enough and galore for me." All photographs taken in service of early psychology of civilization that serviced agrarian demands originating in agricultural thought – quality of attention, significance of place, duality, trophic anxiety attitude towards possessions and domestication.[6]

As a developing nation through the mid- to late-nineteenth century, Australia began to galvanize a sense of its own national character as a separate identity to its colonial pedigree. Early mateship ideologies and bush mythologies, which were reinforced through photography, saw the European settler in contestation with the wild fluctuations of the natural world. To be Australian meant to survive on a disconnected continent inflicted with environmental extremes of fire, flood, storm and drought. Furthermore, these natural fluctuations were concurrent with a sustained confrontation on the frontiers with Australia's indigenous populations. It was a nationhood burgeoning in an ontogeny without ecology. This white-mateship ideology was further reinforced in the early twentieth century with the nation's participation in the First World War, an involvement that many consider vindicated Australia's right-of-passage to nationhood. Imagery of Australian landscape subsequently became imbued with this ideology and nationalistic sentiment. This is no more apparent than in work such as H. J. King's sublime images of Cradle Mountain, or Frank Styant-Browne's photographs of Tasmania's Western Tiers, or Frank Hurley's images from remote central Queensland or the pictorial representations of land distributed as mass-produced postcard images by Richard Daintree; all images which rebuilt the ecological reality of Australian bush in to the "paradise" demanded by their Abrahamic forebears.

THE BIRTH OF NATIONHOOD

This ideal of Australian mateship, which had been forged from colonialism and the First World War, created an enduring nationalistic ethos of land ownership for economic exploitation. This nationalism as culture and civilization invariably finds itself in opposition to nature, and as such, it finds parallels in early psychological theory. Freud writes in *Future of an Illusion* of nature being a kind of rousing, but dangerous, wild: a place where environmental factors act upon human capacities to exist as if with deliberated agency. These factors are best measured by their extremes, which principally challenge human economic, social and political gains as being entirely separate from environmental value. Freud writes:

> The principle task of civilization, its actual raison d'être is to defend us against nature. … But no one is under the illusion that nature has already been vanquished; and few dare hope that she will ever be entirely subjected to man. There are elements which seem to mock at all human control; the earth which quakes and is torn apart and buries all human life and its works; water, which deluges and drowns everything in turmoil; storms, which blow everything before them. … With these forces nature rises up against us, majestic, cruel and

> inexorable; she brings to our mind once more our weakness, and helplessness, which we thought to escape through the work of civilization.[7]

This opposition between civilization and nature finds a psychological match in early bush mythology and early Australian nationalistic sentiment. Freud continues this position in *Civilization and Its Discontents* (1930), stating, "Nature ... is eternally remote. She destroys us – coldly, cruelly, relentlessly." Such descriptions of the natural world's resistance to human perseverance forms part of the dialectic of sublime nature, but equally, picturesque renderings of the natural environment have the capacity to civilize, or perhaps humanize nature. Freud's language is familiarly idiomatic to romantic landscape paintings depicting the inspiring power of nature, such as the works of Turner or Friedrich. Similarly, it speaks to Australian mateship ideology, and further to the formalist American photographs of Ansel Adams and Edward Weston, in which nature is remote and unknowable, but available for aesthetic exploitation. While Freud's psychological language is not describing an aesthetic mode, it nonetheless mirrors colonialist sentiments towards land, which is in turn reflected in an aesthetic actively maintained by many contemporary Australian photographic practitioners and curators. This aesthetic is dissociative in the Anthropocene, because it falsely assumes humans are separate from the environment.

Australian photography remained particularly loyal to this aesthetic, which was essentially nationalistic, well beyond the mid-twentieth century. Australians such as H. J. King – particularly in his latter photographic works – and Jack Thwaites – especially his editorial contribution to the photographic branch of the Tasmanian Lands and Surveys Department, and later as the Superintendent of Scenic Reserves – favoured displaying their country's environment as a sublime, beautiful and untameable wild. Like the American formalists Ansel Adams, Edward Weston and Minor White, they depicted the natural assets of their respective countries with a nationalistic vigour in search of their own Eden. While the sense of place and subsequent attachment to their surroundings of each of these photographers is admirable in its intentions, the result is a landscape valued for its aesthetic merits, not its wider biospheric importance. As a consequence, land is inevitably aestheticized for its consumptive potential, and set aside for touristic and economic exploitation based upon human-centred aesthetic considerations. These considerations, which often led to places being gazetted as national parks due to their natural beauty and subsequent touristic value, such as Yosemite National Park in the United States, are mirrored all over regional Australia. A comparatively less spectacular example is The Rock Nature Reserve in the NSW Riverina, gazetted in the same year and month as its American counterparts, which, while otherwise unusable for agriculture, is often extoled for it aesthetic rather than its environmental or cultural merit.

Despite their naturalist and environmentalist pedigrees, the work of many landscape photographers, while invariably technically proficient, nonetheless still bore an imprint of lingering colonist and agrarian ideologies. These images made the wilderness safe for consumption, and were often disseminated as picture postcards, and later, coffee table books, calendars and related paraphernalia. These objects are in fact nothing more than aesthetic tokens, which skew our perspective of our place within nature, and stand for little more than a superficial reverence for the landscape. Ultimately desuetude, their superfluousness conceals our necessary ongoing connectivity with the natural world. Great landscape photographers such as Ansel Adams are products of the environmental knowledge of their time, and thus cannot be held entirely to account for the commercialization of beautiful spaces. Indeed, Adams would no doubt condemn the current state of Yosemite National Park and its seemingly endless campervan parking spaces, but modern photographers must draw on contemporary knowledge to redress this legacy of environmental conservation.

To continue to photograph only what is beautiful and consumable is to maintain the Freudian analysis of civilization versus nature, which is increasingly problematic as the footprints of anthropogenic climate change become globally encompassing. When nature is depicted as something to be overcome, or given economic or anthropocentric value alone, it serves to deepen our dissociation from the world about us. When nature is seen as some kind of "otherness" to human civilization, we are arguably hindered in adopting positive climate behaviours, instead assuming overconfidence in the capacity of human-made technologies to solve environmental problems. This concept is evident in broad terms, not just photographically, in contemporary discourse about our ongoing relationships to place, and to the environment as a whole. It permeates as a lack of ownership of our lived realities, highlighting a kind of dissociative behavioural problem, in which we are unable to reconcile with the natural world, resulting in a technologically induced eco-trauma. Further, the traditional landscape photograph, as discussed earlier, can manifest as an evidential document of this trauma.

ECO-PHOTOGRAPHY AND POLITICAL ACTIVISM

Freud in later writing makes an important concession in the delineation between civilization and nature in *Civilization and its Discontents*, which many ecopsychologists and environmental philosophers point to as the embryo of an ecology/psychology nexus (and that might also underpin a new photographic way of thinking): "Our present ego-feeling is, therefore, only a shrunken residue of a much more inclusive – indeed, an all-embracing feeling which corresponded to a more intimate bond between the ego and the world about it."[8] Freud was primarily referring here to the somatic trauma of birth; however, this distress has broader interpretations within the wider discourse concerning our ongoing disconnection from the natural rhythms and cycles of life. It is from this period and ultimately emerging as a critical psychological dialogue post-Second World War that environmental research should question whether psychological wellness might itself be principled through a problematic anthropocentric lens. It might be that a current account for good interior and exterior relationships was being modelled upon systems that are themselves, principally environmentally dishonest. It is from this perspective that Erich Fromm came to write in The Sane Society that: "The fact that millions of people share the same vices does not make these vices virtues, the fact they share so many errors does not make the errors truths, and the fact that millions of people share the same form of mental pathology does not make these people sane." In *Technology, Trauma and the Wild*, research-psychologist Chellis Glendinning builds on Fromms' work, and further Paul Shepards' thoughts and links this disconnect between the ego and the world as a techno-addiction, stating, "The hallmark of any addiction is the presence of denial."[9] This is to say that our systems of illustrating psychological wellness in our relationships to the planet might themselves be an act of addiction and denial. Glendinning continues:

> Technological society's dislocation from the only home we have ever known is a traumatic event that has occurred over generations, and that occurs again in each of our childhoods and in our daily lives. In the face of such a breach, symptoms of traumatic stress are no longer the rare event caused by freak accident or battering weather, but the stuff of every man and woman's daily life.[10]

This technological addiction manifesting as trauma highlights a continued alienation between nature and human culture, which Paul Shepard referred to in his study *Nature and Madness*, as an "ontogenetic crippling," which "reaches back to the invention of agriculture – the crucial point at which human culture achieved a false sense of separation from the natural habitat."[11] Much photographic exploration of the relationship between nature and culture stems from a response to the rise of consumerism in post-war America, with equivalent sociological change spreading worldwide. This consumerism was challenged both in terms of psychological need and photographically in our human/nature relationships. The *New Topographics* exhibition in America in 1975 was a key contributor to a new way of depicting urban landscapes in particular, in which the participating photographers avoided the traditional consumptive landscape aesthetic, instead presenting their subjects in a more realistic, accurate, but nonetheless critical manner. This critical approach to the socio-environmental realities of the burgeoning metropolitanism also led ecopsychologist Paul Shepard to write at this time of the crippling effects on children's development through suburbanization that "if we replace the soft earth with pavement, we will learn in our child's heart that the planet is a desert and a dead rock."[12]

Such representation with critical and political undertones is apparent in the work of Australian environmentalists of the same period, such as Olegas Truchanas and Peter Dombrovskis. Indeed, Dombrovskis' photograph titled *Rock Island Bend* (1980), which depicts a picturesque scene in Tasmania's Franklin River that was scheduled to be decimated by the proposed Franklin dam, sparked widespread public interest in protecting the natural environment, and is considered by many to have contributed to the demise of the Fraser Government in the 1983 Australian Federal Election. Photography had realized its capacity for social documentation and commentary, and was now holding a mirror to Australian nationalism, colonialist mythologies, and extractive industries built on unrestrained growth. It would, however, very soon after face further challenge in the increased democratization of imagery that would come with the electronic image and its modes of dispersion.

PHOTOGRAPHY FOR ENVIRONMENTAL MUTUALISM

Contemporary photography has blended with social convention and is now part of everyday life, such that it evidences Australia's ideology, psychology and cultural philosophies. Like agriculture or automobile production, the business of photography still favours the ideological and economic traditions of standardization, efficiency and fragmented linearity. Its obedience to productivity brings with it haste, impatience and competitive extremes. As such, photography now fulfils an economic quota, but cares little for the environments or people whose livelihoods are negatively implicated. Such an economic and commodity driven ideology creates an intervening cultural and social construct that can hamper the mediation of the natural environment through technology.

Regional Australia, in particular, is subject to a neoliberal economic system, which serves a corporate metropolitanism that can obscure environmental consequences and urban decline. Recent works by regional photographer Jamie Holcombe, with his exhibitions *Civic Melancholy* (2010) and *Civic Malaise* (2015), and Jacob Raupach's exhibitions *The Alpine Way* (2013) and *Fell* (2015), help bring the city-centric urban decay of industry to the fore. Holcombe's melancholic images concentrate on formal compositions of sites in a state of transitory urban decay, while Raupach employs juxtaposition of nostalgic documents of business optimism and commercial reality. These images disclose the metropolitical influence over regional environments, while revealing significant shifts in photographic and environmental thinking and arts practice in regional Australia. What each artist brings through social documentary to the fore is, as ecopsychologist Paul Shepherd states, that the "dream of omnipotence is an infantile dream that should diminish, rather than grow, with emotional maturity."[13] Holcombe does this through an ironic positioning of natural and cultural elements inside a single frame, while Raupach teases out perceptions of our environmental legacy through the ironic placement of photographs and objects in a state of pseudo-museology.

Rodney Giblett writes in *Photography and Landscape* that:

> Landscape photography is one of the major ways in which modern technologically savvy people relate to the land and the land is mediated to us. Landscape photography in tourism, conservation and culture has played an important role in forming and maintaining national identity. It has played, and still plays, an important, but misunderstood, role that is not aware of the cultural and environmental politics of pictures.[14]

Artists such as Holcombe and Raupach explore these politics through their photography by documenting the Riverina region of New South Wales: the space in which they live and work. Photographs of the Australian landscape have previously, by and large, functioned to maintain three predominant colonial views: that of land available for endless agricultural or other extractive value, principally to benefit human consumptive growth; that of untouched and undisturbed wilderness set aside for touristic and economic exploitation; or that of pioneer and mateship mythologies, including numerous stories of backing the (mainly white and European) underdog. In a world facing drastic ecosystem, agricultural and economic change, these traditions need to be addressed for their obvious anachronism.

The Riverina's Wagga Wagga-based photographic artist James Farley's recent investigations titled *Lumen Prints* (2015) and *Broken Hill Return* (2014) – made by a symbiosis of natural and photographic processes over time – are an example of a new response to our techno-centric socio-environmental absence. They also form part of a growing movement that includes works like Harry Nankin's *The Impossibility* (2011), Rebecca Najdowski's *Desert Pictures* (2014) and Penny Dunstan's *Rix Creek Rehabilitation* (2015) which uses alternative-photographic processes as allegories for geologic time in order to produce photography that houses evidence of the Anthropocene. The artists of these works employ a camera-less photography, thereby freeing the eye and arms from the constraints of machine production.

In the work of Farley and Najdowski, paper is pre-prepared then, working with rain, drought and fire, left to expose itself to the elements in order to work with the biospheric phenomena. By removing preconceived ideas of documenting the natural and allowing for serendipity, these artists begin to break down modes of authorship and landscape expectations. In the work of Nankin and Dunstan, photograms or shadowgrams are used as allegory for an unknowing of the environmental other.

Christopher Orchard also addresses this discord between urban and agrarian life in his photographic installation works *Eutierria* (2014), *Soliphillia* (2014) and *Significant Roadside Environment Area* (2015 and 2016), as does photographer Thomas Parker with his series *Don't Waste Wagga* (2015) and earlier in Shona Pratt's exhibition *Traumascapes* (2014). Each attempts to address the problem of regional techno-salvation through an emotional sense of place-attachment, knowing that people are more likely to care for places to which they feel connected. By addressing regional concerns through a relationship with place, these artists seek to encourage climate-positive behaviour. Cultural geographer John Brinkerhoff Jackson wrote in *Discovering the Vernacular Landscape* of a loss of meaningful connection with our environment, "Our contacts with the environment are not only brief and infrequent, but scheduled … the natural environment thus becomes the setting of an experience, rather than of experience itself."[15] This has been assisted by the traditional landscape aesthetic, which provides a vicarious experience and a corresponding environmental dislocation. Susan Sontag argues in *On Photography* that photography is "a way of certifying experience, taking photographs is also a way of refusing it – by limiting experience in a search for the photogenic, by converting experience in to an image, a souvenir."[16] Over generations, this practice becomes habitual, which generates an increasing resistance to change. However, this resistance is arguably diminished when the photographer has a keen interest in responding to the surrounding environment through a sense of place, but only when the value of that place also includes critical knowledge and symbiosis of its other-than-human agents. Sontag prophetically warns us of the consequences if we do not grasp this concept: "A photograph is not just the result of an encounter between an event and a photographer; picture taking is an event itself, and one with ever more peremptory rights-to interfere with, to invade, or to ignore whatever is going on. Our very sense of situation is now articulated by the camera's interventions."[17] Photo-theorist Deborah Bright later reinforces this: "Whatever its [the photograph's] aesthetic merits, every representation of landscape is also a record of human values and actions imposed on the land over time."[18] As we experience images through increasingly democratized social and digital platforms, our relationship with the natural world is further disconnected, unless the images can articulate meaningful relationships to place, embedded not as locality, but as experience. If the human animal is not allowed to develop both eco and socio-kinships, this process might result in a kind of picture-viewing or picture-making acting as yet another disease of dislocality. A photograph of land should become more than a landscape photograph, and instead should embedd contemporary social documentation, which traces our relationship with the land. This can result in photographs that depict a more honest representation of regional Australia, which acknowledge the biophysical and psychological impacts of negative environmental change. The photography of Orchard, Pratt and Parker works to try and generate social documents that address the bio-physical and psychological changes happening in regional New South Wales through traditions of documentary, new documentary, post-photography and post-disciplinary making.

Clive Thompson writes in *Global Mourning: How the Next Victim of Climate Change will be our Minds*: "Australia is suffering through its worst dry spell in a millennium. The outback has turned into a dust bowl, crops are dying off at fantastic rates, cities are rationing water, coral reefs are dying, and the agricultural base is evaporating."[19] There is overwhelming scientific consensus that climate-warming trends over the past century are very likely due to human activities.[20] These trends, if they remained unchecked, are likely to contribute to a reduced productivity in rural and regional communities, and an eco-collapse will have significant effects on environmental tourism.[21] In a report titled *Heatwaves: Hotter Longer and More Often*, the (now) independently funded Climate Council stating that Australia's climate has "shifted, and is continuing to shift, increasing the likelihood of more extreme weather."[22] Australian academic and ecopsychologist Glenn Albrecht has documented interviews with scores of Australian farmers over the last decade. He describes their "deep, wrenching sense of loss as they watch the landscape around them change," and cites that "many Australians no longer feel like they know the place they have lived in for decades."[23] Albrecht refers broadly to this state as Solastalgia, a homesickness you feel when you are still at home. I would argue further that a cultural childhood and nationhood deprived of wider earth-value divorces us from meaningful capacities to reconcile problems within an infantile dream of technologic and cultural omnipotence. Albrecht has been developing a psychological dialect that articulates the effects and responses to our dissociative behaviour towards our environment, including the elusive solution to our Solastalgia, namely Soliphilia, which stands for our united love and responsible care for the environment that is our place. There are regional artists, such as those above and many others, who are actively working towards achieving this goal of a combined and sustainable future that build a new nationhood without nature/culture as oppositional forces, a kind of entangled being.

CONCLUSION

It is inadequate to consider the ramifications of Solastalgia in purely economic terms, because to do so promotes anthropocentrism. Photography for environmental mutualism has to reject the traditions of the sublime and the picturesque in order to foster a new national identity, one that is derived in a new cultural, political and economic pluralism for global sustainment. This allows a nexus between photography, psychology and ecology, which can ultimately articulate an aesthetic that does not exclude climate change (and its equal impact upon humans and nonhumans). It also has the potential of bringing new-materialist and post-humanist thinking of nature/culture entanglement in to a whole-of-earth approach to psychological agency that addresses the bioethics and biopolitics of alternate consciousness for inclusive psychic equanimity.

Photography is a cultural document capable of being read as an ideological and a psychological marker. When used with an honest and responsible respect for the environment, it can provide us with a visual ecopsychological impact statement. This reassessment of the division between nature and human culture in the way we document landscape will in turn generate landscape discourse more broadly throughout the creative or cultural industries. It is timely for the establishment of a new multidisciplinary regional centre for environmental education; a centre for the arts and environmental research based in, and responding to the needs of regional Australia; its ecologies, its practices of making, and the relationship of materiality to a renewed discipline of ecopsychology.

NOTES

1. Liz Wells, *Photography: A Critical Introduction 3rd Edition*, 3rd ed. (New York: Routledge, 2008), 46–47.
2. Theodore Roszak, *The Voice of the Earth: An Exploration of Ecopsychology*, 2nd ed. (Grand Rapids: Phanes Press, 2001).
3. Ralph Metzner, "The Psychopathology of the Human-Nature Relationship," in *Ecopsychology: Restoring the Earth, Healing the Mind*, eds. Theodore Roszak, Mary Gomes, and Allen Kanner (New York: Sierra Club, 1995), 55–67.
4. Rod Giblett and Juha Tolonen, *Photography and Landscape* (Bristol: Intellect Books, 2012), 89.
5. Paul Shepard, *Nature and Madness* (Athens: University of Georgia Press, 1982), 37.
6. Ibid., 39.
7. Sigmund Freud, *The Future of an Illusion* (New York: W.W. Norton, 1989), 15.
8. Ibid.
9. Chellis Glendinning, "Technology, Trauma and the Wild," in *Ecopsychology: Restoring the Earth, Healing the Mind*, eds. Theodore Roszak, Mary Gomes, and Allen Kanner (New York: Sierra Club, 1995), 41–54.

10. Ibid.
11. Shepard, *Nature and Madness*, 109–30.
12. Shepard, *Nature and Madness*, 103.
13. Ibid.
14. Sigmund Freud, *Civilization and its Discontents* (New York: W.W. Norton, 1962), 68.
15. John Brinckerhoff Jackson, *Discovering the Vernacular Landscape* (New Haven: Yale University Press, 1986), 63.
16. Susan Sontag, *On Photography* (London: Penguin, 1977), 4–23.
17. Ibid.
18. Deborah Bright, "Of Mother Nature and Marlboro Men," *Exposure* 23:1 (1985): 1–12.
19. Clive Thompson, "Global Mourning: How the Next Victim of Climate Change Will Be Our Minds," *Wired* (January 2008): 70.
20. John Cook, Dana Nuccitelli, Sarah A. Green, Mark Richardson, Bärbel Winkler, Rob Painting, Robert Way, Peter Jacobs, and Andrew Skuce, "Quantifying the Consensus on Anthropogenic Global Warming in the Scientific Literature," *Environmental Research Letters* 8:2 (2013): 1–7.
21. Ross Garnaut, "Climate Change Impacts on Australia," *Garnaut Climate Change Review: Australia in the Global Response to Climate Change* (Melbourne: Cambridge University Press, 2011), 128–33.
22. Lesley Hughes and Will Steffen, *Heatwaves: Hotter, Longer, More Often* (Sydney: Climate Council Press, 2014).
23. Thompson, "Global Mourning," 70.

BIBLIOGRAPHY

Bright, Deborah. "Of Mother Nature and Marlboro Men." *Exposure* 23:1 (1985): 1–12.

Cook, John, Dana Nuccitelli, Sarah A. Green, Mark Richardson, Bärbel Winkler, Rob Painting, Robert Way, Peter Jacobs, and Andrew Skuce. "Quantifying the Consensus on Anthropogenic Global Warming in the Scientific Literature." *Environmental Research Letters* 8:2 (2013): 1–7.

Freud, Sigmund. *Civilization and its Discontents*. New York: W.W. Norton, 1962.

———. *The Future of an Illusion*. New York: W.W. Norton, 1989.

Garnaut, Ross. "Climate Change Impacts on Australia." *Garnaut Climate Change Review: Australia in the Global Response to Climate Change*. Melbourne: Cambridge University Press, 2011, 128–33.

Giblett, Rod and Juha Tolonen. *Photography and Landscape*. Bristol: Intellect Books, 2012.

Glendinning, Chellis. "Technology, Trauma and the Wild." *Ecopsychology: Restoring the Earth, Healing the Mind*. Eds. Theodore Roszak, Mary Gomes, and Allen Kanner. New York: Sierra Club, 1995, 41–54.

Hughes, Lesley and Will Steffen. *Heatwaves: Hotter, Longer, More Often*. Sydney: Climate Council Press, 2014.

Jackson, John Brinckerhoff. *Discovering the Vernacular Landscape*. New Haven: Yale University Press, 1986.

Metzner, Ralph. "The Psychopathology of the Human-Nature Relationship." *Ecopsychology: Restoring the Earth, Healing the Mind*. Eds. Theodore Roszak, Mary Gomes, and Allen Kanner. New York: Sierra Club, 1995, 55–67.

Roszak, Theodore. *The Voice of the Earth: An Exploration of Ecopsychology*, 2nd ed. Grand Rapids: Phanes Press, 2001.

Shepard, Paul. *Nature and Madness*. Athens: University of Georgia Press, 1982.

Sontag, Susan. *On Photography*. London: Penguin, 1977, 4–23.

Thompson, Clive. "Global Mourning: How the Next Victim of Climate Change Will Be Our Minds." *Wired* (January 2008).

Wells, Liz. *Photography: A Critical Introduction 3rd Edition*, 3rd ed. New York: Routledge, 2008.

Chapter 10

Nature as Creative Catalyst

Building Poetic Environmental Narratives through the Artists in Antarctica Programme

Octavia Cade

Antarctica is a continent of extreme and varied geography, but such is its latitude that many do not find a point of close identification with even one of its environments. Unfamiliar landscapes can be perceived as difficult to understand, especially if they are associated in the minds of laypeople with science or controversy. The beauty of Antarctica, for instance, can be shared via coffee book tables or gallery photographs of vast landscapes and leopard seals, but communicating climate change and its effect on ice shelves can be more challenging.

More than all others, Antarctica is a continent of science. The majority of its human inhabitants – all of whom are temporary residents, with an often seasonal presence – are attached to research stations devoted to exploring natural history through reasoned and methodical investigation of the surrounding environment. This necessarily has an effect on the prevailing experience of the natural: it becomes something to be dissected, and yet the methods of dissection are such that the entanglements occurring on the continent are complex, documented and limited. Because the continent is so distant, and so inhospitable to human life, there developed in Antarctica an ecosystem that was in many ways *disentangled* from human processes and phenomena. It is only in the last century or so that direct entanglement has begun to be observed – and, crucially, documented. This is different from the human entanglement with ecosystem that has taken place in other, more hospitable areas of the world. There, habitation often spans prehistoric time scales, but because Antarctica has been explored – and through science, colonized – so late in human history, the ability to document entanglement (by explorers, by photographers, by naturalists and scientists, by artists and writers) has been present from the beginning.

This entanglement occurs in a number of ways, and not always directly. Increased population and industrialization in the rest of the world has had subsequent consequences on the physical state of Antarctica, for instance the effects of climate change on ice shelves, or the transportation of microplastics to Antarctic shores through the mechanism of ocean currents. Yet ecological evidence of entanglement is also apparent in the establishment of permanently staffed research bases. These bases have, for the first time, established long-term, local human communities on the continent. These localized entities are now part of the polar ecosystem, part of an ongoing colonizing process, and the subsequent entanglements are spreading – if reluctantly, and often under severe state restriction.

Much of the perceived scientific and natural value of Antarctica, after all, stems from its isolation from human influence, yet the very acts of observation and testing require human presence in – and encroachment on – the original ecosystem. Entanglement – especially on a direct, local scale – therefore occurs whether it is wanted or not, because it is impossible for the ecosystem *not* to change with the introduction of a new element. The challenge lies in limiting human effects on the Antarctic ecosystem as much as possible. But by limiting visitors to (primarily) scientists, human understanding of that ecosystem risks equal limitation.

When scientists are the primary workers in any given natural environment, it is *their* experience of that natural that will predominate, and *their* language and texts that will communicate their experiences and values, for the two are not isolated things. As Barad notes when talking about "entanglements of facts and values … values as well as facts are being 'cooked' together at the lab bench,"[1] and it is unlikely that the lab bench situated in the Antarctic will prove to be the exception. Yet even scientists will admit that scientific papers are not welcoming documents with which to transfer their findings and beliefs. Different methods are needed if the wider society is to participate and learn about Antarctic life and environment.

NARRATIVE AND NATURE

Narrative provides a familiar, unthreatening framework that is often more successful in illustrating natural history than tables of fact, which are frequently presented in unflattering or unmemorable forms. Narrative, however, is a competitive phenomenon. There is no single "Antarctic narrative," and no such narrative that is not informed by others already in existence. Human engagement with the natural environment, for instance, is frequently complicated and shaped by distance. Because direct experience of the Antarctic environment is so very limited – generally restricted to explorers or scientists – distance allows for the mythologizing of the continent. Narratives are built-up second-hand, through the experiences of others, and these experiences often privilege the historic, or the exciting singular event. Such emphasis is only reinforced by the diaries and other records of the Heroic-Era Antarctic explorers, whose texts tell stories of privation and isolation and the occasional nervous breakdown. While these writings may have been a way to "maintain sanity" in a hostile environment,[2] they also contribute to the shaping of a narrative of a hostile nature; one where human settlement is a fragile and precarious thing, an environment fit only for heroes and martyrs. The expeditions of Scott and Shackleton, the disaster of the Erebus air crash, all help to build an environmental narrative of danger and hardship that is only emphasized by distance, by travel restrictions and governmental reserves, and by the difficulty of rescue.

This narrative can be reinforced in a number of different ways. Two of these are illustrated in Bill Sewell's book *Erebus: A Poem*. In 1979, Air New Zealand flight 901, on a polar sightseeing trip out of Auckland, crashed into Mount Erebus, on Ross Island, Antarctica. All the 257 passengers and crew were killed, and the subsequent, highly controversial investigation went all the way to the Privy Council. Sewell uses primary sources to inform his poems, interspersing his own verse with excerpts from that investigation. Most affecting is the poem "Frame 2: The transcript,"[3] which includes sections from the cockpit voice recorder which was recovered after the crash. The transcript of the pilots' voices, the confusion at their location transforming into terror as the volcano looms suddenly before them, their failed efforts at survival, underlines the collection as a factual document, as something based on actual rather than imagined disaster. The poem is almost uncomfortably voyeuristic; it is hard to read it and not imagine the tone and panic of the recorded voices.

Adding to the verisimilitude provided by the documentary evidence are the notes collected at the conclusion of the collection.[4] Endnotes are a typical feature of academic or non-fiction texts: although they are not unknown in poetry, they are less usual as a feature of creative work. It is the dreadful, painful knowledge in them that informs the poems, however, and their relationship with both the Erebus disaster and the natural environment of the Antarctic. The Erebus volcano, one endnote tells the reader, was so named by Sir James Ross after one of his two Antarctic polar ships. The other ship, the alternative not used, was the *Terror*. Naming the volcano as he did in 1841, Ross can hardly be expected to show prophetic ability, yet given the inclusion of the pilot transcript in the "Frame 2" poem, readers can hardly be blamed for thinking that Terror would have been a more appropriate name for the volcano in question.

The advantage of using such well-known events, of such well-used narratives, is that it relies on a general knowledge already possessed by much of the reading audience. The cultural and historic associations with the names "Shackleton" or "Erebus" are such that an author may rely upon the shared experience of that audience to inform his or her work. This mutual understanding provides an anchor with which new information, and new associations, can be inserted into the text without the reader feeling either lost or overloaded. It is a shared context that not only informs but also deepens the associations, the narratives, behind the work. The more that Antarctica is associated with the Erebus crash, for example, the more entrenched the narrative of "Antarctic disaster" becomes in the minds of the readers. Such is the risk and reward of context: it informs but it also entrenches – often at the expense of other narratives. But context can also be individualized, and the resulting narratives are subtler and smaller, more diverse, and informed by the personal rather than the mythic. This allows individuals who work with nature to offer up their own experiences, which frequently move counter to (or even wholly independently of) the most popular narratives.

ARTISTS IN ANTARCTICA

This diverse engagement with the natural environment, and how personality and individual context informs the resulting environmental narrative, is illustrated in New Zealand's Artists in Antarctica programme. This has, in each year since the inaugural 1997/1998 season, sent artists to Scott Base, the New Zealand research base in Antarctica. Located on Ross Island in New Zealand's Ross Dependency, Scott Base supports up to 85 people at a time. While some of the residents are there for logistical purposes, many are scientists. Projects currently being undertaken in the 2015/2016 season include research on Adelie penguins (*Pygoscelis adeliae*), Antarctic toothfish (*Dissostichus mawsoni*), supercooled sea water, sea ice, and dry valley ecosystem resilience.

The Artists in Antarctica spend time on the ice in order to interpret it creatively for public engagement. While they are not officially attached to any of the specific research programmes, they are able to spend time with (and sometimes shadow) the participating scientists: for instance, on a trip to an Adelie penguin colony. The aim of this interdisciplinary, collaborative programme is to promote the values and the understanding of the Antarctic, from both scientific and aesthetic perspectives (a similar scheme – The Antarctic Visiting Artists and Writers Program – has been offered, since 1957, by the American National Science Foundation). The artists that have gone to the Antarctic under this New Zealand programme are diverse: they include visual artists, musicians and filmmakers as well as writers. Particularly telling are the collective experiences of the poets, whose reactions to a common experience yield a wide range of narratives about the southernmost of our natural environments. Paradoxically, this diversity is best illustrated in instances of commonalities of experience. When multiple poets have the same scientific, environmental or natural experience, the stories of their reactions to that experience – even if each reaction is split between individual poems – will be consistently different. Each will produce a fragmented narrative, where individual poems link together to form a broader picture, a story of how the author engages with his or her environment, and that narrative can be used to inform the reader. Poets who experience widely different environments and ecologies can make for difficult comparisons: they may, for instance, share a focus in environmental activism, but the challenges, conflicts and issues facing the Great Barrier Reef differ from those of the American grasslands, which differ in turn from polar glaciers. Attempts to tease out similar approaches, similar impressions, are made more difficult by the fundamental differences in the natural subject. Indeed, the same can be said for other artists as well: sculptors, painters, textile workers and filmmakers, cannot easily be extricated from the natural environment of their subjects.

In comparison, the various artists and writers who visit the ice under the direction of Artists in Antarctica have shared, in different years, a similar experience. They have all travelled from Christchurch to Scott Base and back; they have all stayed at the Base and accompanied scientists about their work, experiencing the landscape, inhabitants and research programmes that the new environment has to offer. They also exist in a position of half-privilege between two worlds: while they are government-funded artists advantaged by access not granted to their peers, the Antarctic artists as a group can nonetheless be perceived as interlopers into the scientific, irrelevant and effectively second-class polar citizens[5] whose values and experiences do not match those of the more firmly entangled scientists.

These shared aspects culminate in a remarkable commonality of experience between the attending poets, and yet the resulting narratives produced by them are quite different.

Of course, the narratives produced by the Artist in Antarctica poets were themselves informed by existing Antarctic narratives. In his edited collection *The Wide White Page: Writers Imagine Antarctica* (2004), Manhire comments on the effect of previously imagined stories: "Beyond the realms of climate and geology, the early history of Antarctica is a tale of the centuries-long search for the place which bore that name – or for that overlapping unknown continent, *Terra Australis Incognita*."[6] Stories no more emerge from a vacuum than language does, and true science or natural history stories emerge from a primordial narrative soup of fact and fiction combined, of false leads and blind groping. Many of these earlier Antarctic works were shaped as much by ignorance as imagination. But "as the southern ocean was actually entered and traversed, Antarctic narratives began to report authenticating detail" from scientific records, from the reports brought back by explorers.[7] Manhire argues that this incompletion was a storyteller's happy gift: Antarctica "offered a generous, flexible location – geographically beyond reproach, but still sufficiently uncharted to be able to host any number of narrative surprises."[8] It's this flexibility, these surprises, which can be seen in the poetic output of the Artists in Antarctica programme.

Bernadette Hall's (1945–) experience of Antarctica is full of life and colour and her poems reflect this. Helter-skelter, full of disparate images all melded together, fragments of history and experience, image and humour, the poems are effectively snapshots of wonder. "Has anyone got a camera?" Hall asks in "How Ponting Thought He'd Died and Gone to Heaven,"[9] and it is hard to read Hall's poems and not picture her spinning around in delight trying to take everything in. Hall's text shares a story of almost hyperactive exhilaration. "I'm fading out again in iced excitement," she confides in "The White Boudoir,"[10] nearly drunk with the need to cram in as much novelty as possible, to try and interpret the new world around her. "Like you, I'm trying to read / the *tabula scripta* of the big berg."[11] The presence of ice shelves, of bergs and glaciers, is clearly a novelty. The scale of difference between the Antarctic environment and that of the New Zealand mainland is enormous: temperature, aridity and albedo are all immediately alien, and Hall's previous environmental context is insufficient for translation. In "trying to read" the bergs, Hall is grappling with the constant, overpowering presence of ice.

Crammed as they are, the initial perception of fragments pasted all together fades as the collection takes shape, the poems fitting together to produce a narrative of excitement, of extreme enchantment. This perhaps most apparent in "Door," a thirty-seven line prose poem that is a single breathless sentence.[12] Hall is in a corridor in Scott Base, and the stream of consciousness that results spins and cycles through numerous allusions (wind jackets, Sir Ed, the threats of the base chef, a book of poems by Wallace Stevens, the inability of prose writers to understand figurative language). On and on it goes, all crammed together in one madcap rush. Hall seems to take particular delight in the mashing together of differences and opposites in this way, a tendency that is also well illustrated in another poem of hers, "Millennia of silence and no rain."[13] In this poem, Hall and a young Korean woman, a science technician also visiting the Base, are in the Chapel of Snows. The technician has limited English, and Hall is trying to give a brief lesson in Latin to a woman who "works with weather balloons, translating the news from the galaxy. I'm wondering if she writes it in Korean but maybe it's just numbers."[14]

This isn't a narrative exclusive to Hall's experience of Antarctica. As different poets bring different context to their experiences, so their narratives will be shaped by that context. This open embrace of wonder is something that can be seen in other of Hall's collections. The opening poem of *The Lustre Jug*, "Rathcoola Rain," has Hall again transplanted into a different environment – this time Ireland, for in 2007 Hall spent six months there on the Rathcoola Fellowship. Coming from Antipodean environments characterized by drought and fire and trees whose germination is triggered by such, Hall is struck by the Irish weather: "We walk out into it, we lift our arms up, / we hug the rain."[15] Hall even pokes fun at her own capacity for astonishment. When her daughter briefly fools her into thinking that "gullible" has been removed from the dictionary in "A Writer's Life, or, A Sackfull of Spuds," Hall comments that "I'm amazed and rather ashamed of how easily / I'm amazed and always have been."[16]

Owen Marshall (1941–) has arguably produced the most fragmented narrative of Antarctica. While other poets tend to clump their Antarctic-inspired poetry into a cohesive subsection within a collection, Marshall, in his book *Sleepwalking in Antarctica and Other Poems* (2010), intersperses his Antarctic poems throughout the rest of the text. They are not confined, and they are a distinct minority: the "Other Poems" of the title predominate. Yet the first two lines of the first poem, the titular "Sleepwalking in Antarctica," indicate that this is a deliberate decision. "Four to a bunk room here. / Close familiarity of strangers."[17] Marshall sorts his poems so that the Antarctic verses are cheek-by-jowl with images of bike riding and tussocks and poplars, with Norfolk Island and ruins at Mycenae. This experience of dislocation and disjunction is repeated in the other Antarctic poems. "Snow Poem 3: Metaphysical Pole" comments that "On a pillar in the McMurdo bar / was written, 'metaphysical pole'"[18] while "Snow Poem 10: Flags" comments on the multitudes of colourful cloth that "They are a minor but defiant intrusion on / the world of black and white."[19] It is difficult to read Marshall's experience without being struck by his feel for contrast. The image of the flags is particularly telling, illustrating a tension between introduced and natural elements in a Base that is itself an intrusion in a once pristine landscape. It is a landscape defined by limited palette. Marshall refers to it as a "world of black and white," defined essentially through the colours of ice and snow and rock, and the physical world reflects in the biological. Penguins, seals, birds are species that have evolved to the Antarctic environment, that have adapted in their colouring. The brightness of the intruding flags – so noticeable to Marshall – is an artefact of an introduced (human) species, an indication of dislocation, of disruption. It's no wonder that his narrative is fragmented. As with Hall, this fragmentation is something that can be observed in Marshall's other poetry. His collection *Occasional: 50 Poems*, for instance, is equally interspersed in setting, with a poem "At Pergamum"[20] cheek by jowl with one on "Marlborough,"[21] and a comic poem on wine, "Grape iwi"[22] snuggled up to more sombre poem on the death of the solar system, "Relativity and the solar system."[23]

In contrast, Bill Manhire's (1946–) "Antarctic Field Notes" (the central section of his 1999 book *What to Call Your Child*, later reprinted in *Collected Poems*) describe an experience primarily characterized by communication and by interaction. As Marshall's opening poem sets the tone for his fractured narrative, so does "Into the Ice Shelf" set the tone for Manhire's. As the first lines show, the poem reads as a conversation, almost a transcription, of the boat journey into Antarctica.[24]

Right ten degrees rudder!

Right ten degrees rudder, aye!
My rudder is right ten degrees.

Very well.
Ease your rudder.

Ease my rudder, aye.

This narrative of interaction lasts throughout the section. Manhire's experience is that of making connections – is that of the *act* of making connections. His is a less passive story than Marshall's, a story of active engagement with an alien landscape. He is greedy, almost, in his desire to bond with it, as illustrated in his "scientist's love song" for Lake Bonney.[25] "Patch me out to Lake Bonney, / Patch me out to the ice," he says, in his poem "Deep Field Song." "And patch me out to McMurdo," "But then patch me right back to Lake Bonney, / Patch me whatever the price."[26] Manhire's experience prioritizes interactivity, and because of this he is able to note his own effect on the landscape, an effect beyond the ephemera of flags, as he does in "Some Frames."[27]

Antarctica!
where a single
footprint lasts
a thousand years

and here we are
with our
thousand footsteps
etcetera

Because the Antarctic environment is so dry, it is not subject to the same erosion processes as the more liquid continents. There is wind, of course, and winds at the Antarctic can reach enormous speeds, but wind itself is not enough to disrupt the evidence of human activity. One is reminded here of astronauts' footprints on the moon, and Manhire is not the only poet to draw attention to other planets (Orsman also does, as described below). The alien nature of the Antarctic natural environment, where even common physical processes are absent, is underlined in poems such as "Some Frames." The aridity is both comprehensive and characteristic; yet because this desert is built of frozen water there is the potential for disjunction, for the questioning of expected imagery. But questions, by definition, demand answers.

Of course, all poetry is characterized by communication, by interaction – even if, at its most base form, the interaction and the communication takes place between poet and reader. This is a particularly relevant theme, however, when a poet is transplanted into an environment that is so extraordinarily alien to them. Manhire describes that environment as "seriously weird and marginal, real edge-of-the-galaxy stuff. The place feels like a station on the moon or Mars, with only the breathing apparatus missing."[28] But even with the absence of breathing gear, there's another way to commune, and that's with the scientists who live and work at Antarctica. Because the station is so isolated, and science so prioritized, the demographics and culture of the polar station is markedly different from what the visiting poets experience in their normal daily life. Manhire comments, "Scientists can seem like embarrassing, marginal figures. ... But down in Antarctica, they thoroughly matter."[29] He describes the pleasure they take in their (often strange) research, and the affection they have for the environment. That is the point of communication there: the scientists have made a sort of secondary home on the polar continent; and that through them others can come to understand what it is like to live there. Yet Manhire illustrates this "communications" perspective in some of his other poems, although these often don't have the same concentrated sense of immediacy (though this can only be a subjective assessment). His biographical poem on the astronomer John Herschel, "Herschel at the Cape," describes the four years that this scientific polymath spent in South Africa from 1834–1838.[30] While Herschel's primary scientific role was to survey the southern skies, he also applied his talents in other ways. Manhire shows him drawing, hunting, bird-watching: Herschel is actively engaged with his environment in the same way that Manhire was in Antarctica. "Herschel at the Cape" is written in the third person, however, which may contribute to the lesser sense of immediacy than that which Manhire displays in his Antarctic poems.

Similarly, Manhire's use of communication can be seen in his "Erebus Voices," written for the 25th anniversary of the Erebus disaster and read by Sir Edmund Hillary in a commemoration at Scott's Base.[31] "Erebus Voices" consists of two short poems, describing the disaster from the perspective of first the mountain, and then the dead. These are not third person poems: the mountain is "I," the dead are "we": Manhire is essentially giving them voices. Earlier in the *Lifted* collection, he does the same for the polar explorer Roald Amundsen, lifting an experience from Amundsen's own writings (the shooting deaths of many of the sled dogs described in *The South Pole*) and reinterpreting it through verse. Amundsen himself aimed for poetry – "Dogs" is preceded by a quote from the explorer himself: "I tried to work up a little poetry – 'the ever-restless spirit of man' – 'the mysterious, awe-inspiring wilderness of ice' – but it was no good; I suppose it was too early in the morning."[32] This is therefore similar to the Herschel poem in that it is based on a historical personage, but "Dogs" is written in the first person, which allows Amundsen's mixed feelings about the slaughter of his animals to come through loud and clear: "I feel quite alone. It is hard almost to speak … / My best friends bark in my stomach …," he says but continues on, unable (and unwilling) to stop exploring, so fascinated is he by the landscape.[33]

A similar focus on interaction can be seen in the poems of Nigel Brown (1949–). Given that Brown and Manhire were Artists in Antarctica during the same season, it is not surprising that their experiences – and thus their narratives – are closer than the usual. Although Brown is primarily a visual artist, four of his poems can be found in the chapbook *Homelight*, which also includes poems by Manhire and Chris Orsman (1955–). Even with such a small sample size, Brown's grasp of interaction is broad. Not only does he refer to the need for minimizing the ecological impact of human presence in Antarctica (in "Orsman's Cigar"), he also makes comical comment on a (literal) audience of Adelie penguins (in "On the Prospect of Manhire Reading to Penguins"), but also imagines the landscape as pining for human contact, for the interaction between environment and artist (in "Blood Falls Lament"). It is interesting to note here the small pamphlet *Southern Convergence*, which contains (among other things) early sketches and woodcuts by Brown of his time in Antarctica. The "Cape Evans: woodcut" has a small note in verse form, describing the bleakness of the polar environment, but this note ends cheekily, saying that when Brown and Orsman and Manhire visited Cape Evans history asked why they had taken so long to arrive.[34] This sense of age, of time passing, is present in all of the surveyed poets. Partly this can be said to be due to the environment itself: the unchanging, frozen nature of the continent giving an unshakeable impression of a land frozen in time. And yet no one visiting a scientific research station on Antarctica is immune to the *testing* of time. The taking of ice cores, for example, samples atmospheric air particles for periods spanning many thousands of years. The ice itself is ancient, and carries echoes.

The question "*Why did you take so long to get here*" is one that raises an interesting point: the tendency of the poets – and perhaps of people in general – to anthropomorphosize the continent, to ascribe it personality and the capacity for communication and choice. This isn't hard to do when the "other" is a living creature, such as a penguin or a leopard seal, which can easily be seen to exhibit character (in, for example, playful behaviour or by interaction with the other members of their species). It is more difficult to justify the belief that a continent has a similar character, and yet this is often implied in some of the Antarctic poems. The continent asks questions, it has mountains referred to in the first person, it threatens, it preserves. Of course ecosystems have been treated as entities before – James Lovelock's *Gaia* representation of the planetary ecosystem, for one – but there is a temptation to conflate character with agency; with the ability to choose. Does the Antarctic truly exhibit agency by preserving a footprint for a thousand years, for example, or can this poem simply be read as an example of entanglement, and of *human* agency, preserved for millennia by the physical processes of a polar environment? Manhire doesn't come down either way, but readers' interpretations can differ. The entanglements here are complex, and not just ecologically. Literature and history, the social pressures of expectation, the perception of nature all combine to alter the effects of human behaviour on the environment.

This history may be personal as well as communal. Perhaps the most interesting example of New Zealand Antarctic poetic narrative can be seen by comparing two texts by Chris Orsman. *South: An Antarctic Journey* (1996) was written prior to Orsman's own experience as an Artist in Antarctica, and is primarily a historic retelling of Robert Falcon Scott's final, terrible expedition wherein he died of exposure in March of 1912.[35] This is contrasted with *The Lakes of Mars* (2008), written after Orsman had spent time on Antarctica himself, after he had had his own direct experiences.[36] *South* has an appealing framework. It is bookended by short prose pieces about Orsman's grandfather, who as a child in New Zealand had won the prize of a short boat trip on the *Terra Nova* with Captain Scott, before the latter's departure south. This short, framing narrative links the personal to the historical, and introduces the idea of competing narratives. Not only is the writer aware of the voice and story of Scott – fragments from whose diaries appear throughout the text – but also of members of his expedition, such as the photographer Ponting – as well as Orsman's grandfather, and Orsman himself. This is a personal story as much as anything else, the story of a grandson's attempt to connect with his grandfather through a shared interest, a mutual fascination.

The Lakes of Mars, by contrast, is personal in a different way. Direct experience lessens the impact of historic narrative, and Orsman's perception of that narrative is altered. While *South* perceived Scott's story in largely biographical terms, *The Lakes of Mars* removes the sense of continuum from that biography. Scott is also present in the second collection, but he is present almost in cross-section, preserved in ice if not in amber. "The Book of the Dead" tells of Orsman's visit to Scott's hut: "I sniff the air and stand / wide-eyed with the Egyptologists / in a flash-lit chamber."[37] Much of the poem consists of Orsman listing the objects he finds there, and much of the pathos of the poem comes from the inherent unsuitability of these objects[38]:

Thus: a long centre table
half-set with blue-and-white china,
green butter dish, tin plates.
seven patched Windsor chairs.

This is preservation again. The hut is an environment frozen in time, frozen by virtue of its intrusion into an alien environment. The butter dish and tin plates can be seen as analogues for prehistoric air bubbles, trapped in the ice as a sort of time capsule. It is also entanglement of a very particular recursive sort. The presence of the hut and its accompanying artefacts affect the local environment and the physical micro-processes of the landscape (e.g. disruption of wind patterns and the effects of hut construction on soil and erosion). But the preservation of this *initial* preservation is a process in itself. Without ongoing restorative work, this time capsule hut would eventually be destroyed by the physical stressors of the environment. The perceived historical value of the huts demands further entanglement, and the continued disruption of the existing ecosystem.

The realization of Antarctica's inherent strangeness is signposted in the title of the collection, as Orsman opens his series of Antarctic poems with a quote from the environmental historian Stephen Pyne: "The microbiosphere of Antarctica has more in common with Mars than with Earth." This realization is apparent in *The Lakes of Mars* far more than in *South*, and it is a realization that becomes more solid to Orsman through his own experience. His story of Antarctica changes from a personally informed, biographical narrative to a personal encounter with the alien. For Orsman, stepping into Antarctica is like stepping into another world, even the mummified seal he spots in "The Lakes of Mars" has "found her spacesuit lacking"[39] and disintegrates in "the exobiology of this / open-air refrigerator."[40] This experience is compounded by Orsman's feelings of stepping into another time: not just the Edwardian remains of Scott's hut, but the ancient, undisturbed landscape of "waters / that carry the imprint of BC / in their toxic depths,"[41] waters that were "buried alive / when Christ's wound / was fresh in his side."[42]

It is the physical characteristics of the Antarctic environment – the aridity, the temperature, the long seasons – that particularly impress the poets. Primarily, one imagines, because these things are so unfamiliar. Antarctic wildlife may be impressive, but a penguin outside its natural environment is a familiar thing. Many people have seen them in zoos, in films, or even had them in the form of stuffed toys. Yet one cannot cuddle a glacier, or ascribe personality to an ice shelf. It is only in the personal experience of that alien environment – one noted as being more akin to Mars, or to the moon – that true understanding of that environment is able to emerge.

WORKING WITH A CHANGING NARRATIVE

This collective, creative engagement with nature, the use of environment as imaginative catalyst, shows how artists can work with nature to produce narratives that both inform and explore the idea of distant ecologies, and foreign ones. How they observe a natural environment that is primarily the purview of disciplines not their own is another way of working with nature: one that synthesizes multiple observations, one that is informed by interdisciplinary behaviour while not weighed down by it. The diversity of the output after the commonality of experience shows that working with nature is as much a product of personal context, of character and interest, as it is objective methodology and the rigid protocols of science. Narrative is therefore not just a tool of engaging with nature. It is a catalyst for creative endeavour, promoting not just diversity of experience but the *communication* of that experience to others – even if that narrative is of non-traditional form.

Such narratives may be simplistic or complex, but their primary importance lies in the familiarity of their structure. This is especially relevant when communicating both environmental science and poetry, as readers unfamiliar with these disciplines can find them difficult to navigate. Given how closely science is linked with the natural environment of the Antarctic, the scientific language is in many ways the primary language of that continent; the way that the majority of human inhabitants of that continent best express what they find there. Yet because the scientific language is extremely precise, the resulting jargon can be tricky to understand. In contrast, poetic language often relies on layered meanings and metaphors, which can be frustrating to those more accustomed to plain speech. Poetry about science, or poetry about the natural world which is primarily interpreted through science, does not naturally become any more accessible by synthesis of two linguistic difficulties. Narrative, however, is an extremely accessible form of communication, and by using it as a framework the Antarctic poets are able to make their poems more accessible to a general, non-specialized audience.

The narrative expectations of that audience, however, are not always consistent. They may even change over time. The prevailing narrative of Antarctica as the pristine wilderness of Shackleton and Scott, a place where humans can barely survive, can be undermined by modern imagery. An internet survey performed by Summerson and Bishop presented photographic images of Antarctica in an attempt to explore the aesthetic and wilderness values associated with that continent.[43] It is unsurprising that those images containing human figures or evidence of human activity scored lower among respondents: the images were perceived as less aesthetically valuable and less indicative of wilderness. Recall here the many human details observed by the Antarctic poets: the many coloured flags, the butter dish, the china set. These are not images that are immediately brought to mind when one thinks "wilderness," regardless of how elastic that environmental definition may be. Yet the *stories* behind the objects both merge and diverge with that initial influential narrative. The historic huts of the Antarctic only exist because of their role in the story of the exploration of that continent; yet the isolation-narrative so common to many of those stories is undermined by the homely imagery not only of historic dining tables, but of the many modern dwellings now in use on the ice. Research stations such as Scott Base are both isolated and not – geographically cut off, they are also linked to the wider world via internet, by the small reminiscent touches of the resident scientists, who bring their own keepsakes from other lands and other continents, their own small stories and objects.

The opening up of the Antarctic continent to scientists rather than explorers has had a further narrative effect. Roscoe's 1958 survey of Antarctic literature notes that the history of exploration on that continent is often biased towards adventure stories. These privilege the previously mentioned narrative of danger and excitement; Roscoe notes that, in one particular case, "History suffers, but the spirit of the Antarctic is there."[44] Adventure narratives remain, of course. Not only in periodic explorations and recreations thereof, but in the experiences of the artists themselves: no-one could read Bernadette Hall's Antarctic poetry, for instance, without believing she found herself to be on a grand adventure. But as Antarctica became more accessible, with greater technological and logistic ability allowing for permanent habitable bases, the narrative ratio began to change – from the explorative to the scientific. Poets follow scientists, filmmakers follow penguins. The exploration primarily comes now garbed in the clothes and habits of science; the natural world taking on the work-shape of laboratory instead of frontier.

CONCLUSION

This is a natural consequence of changing patterns of visitation. Nicol argues that "Antarctica, more than most destinations, is … defined by the perspective of the visitor."[45] When visitors are so comparatively few, their accounts gain extra weight and the creative engagement of those visitors with the continent gains extra privilege thereby. It should be remembered, however, that a frozen continent is not a rigid one. That Antarctica is now primarily a continent of science, of workers who engage in a methodical investigation of the natural environment, does not mean that theirs is the default position forever more. Climate prohibits general settlement, and science will always have a primary position in the future of the Antarctic continent, but increased access increases diversity of visitors. A thriving tourist trade exists, and the importance of opening up the continent to wider, more engaging discourse than that of the scientific is acknowledged as being of increasing importance. Indeed the Antarctic Treaty respecting the international cooperation in and governance of that continent refers to the fact that Antarctica has "been the subject of significant works of art, literature and music" and that promotion of the "scientific, aesthetic and wilderness values" should be promoted through the "contribution of Writers, artists and musicians."[46] Given this, one can only wonder what the poets will make of Antarctica in the future, and what stories they will bring back with them.

All quotes from the works of Bernadette Hall, Bill Manhire, Owen Marshall and Chris Orsman used with permission.

NOTES

1. Malou Juelskjær and Nete Schwennesen, "Intra-Active Entanglements – An Interview with Karen Barad," *Kvinder, Køn & Forskning* 12 (2012): 15.
2. Elizabeth Leane, "Isolation, Connectedness and the Uses of Text in Heroic-Era Antarctica: The Cases of Inexpressible and Elephant Islands," *Island Studies Journal* 2, no. 1 (2007): 67.
3. Bill Sewell, *Erebus: A Poem* (Christchurch: Hazard Press, 1999), lines 15–16.
4. Ibid., line 61.
5. Elena Glasberg, "Camera Artists in Antarctica," *New Zealand Journal of Photography* 65 (2007): 23.
6. Bill Manhire, "Introduction," in *The Wide White Page: Writers Imagine Antarctica*, ed. Bill Manhire (Wellington: Victoria University Press, 2004), 9.
7. Ibid., 11.
8. Ibid., 12.
9. Bernadette Hall, "How Ponting Thought He'd Died and Gone to Heaven," *Life & Customs* (Wellington: Victoria University Press, 2013), line 10.
10. Hall, "The White Boudoir," *Life & Customs*, line 11.
11. Hall, "Le Monde Sauvage," *Life & Customs*, line 11.
12. Hall, "Door," *Life & Customs*, 64–65.
13. Ibid., 40.
14. Ibid., lines 8–10.
15. Bernadette Hall, "Rathcoola Rain," *The Lustre Jug* (Wellington: Victoria University Press, 2009), lines 6–7.
16. Hall, "A Writer's Life, or, A Sackfull of Spuds," *The Lustre Jug*, lines 7–8.

17. Owen Marshall, “Sleepwalking in Antarctica,” *Sleepwalking in Antarctica and Other Poems* (Christchurch: Canterbury University Press, 2010), lines 1–2.

18. Marshall, “Snow Poem 3: Metaphysical Pole,” *Sleepwalking in Antarctica and Other Poems*, lines 1–2.

19. Ibid., lines 6–7.

20. Owen Marshall, “At Pergamum,” *Occasional: 50 Poems* (Christchurch: Hazard Press, 2004), 36.

21. Marshall, “Marlborough,” *Occasional: 50 Poems*, 37.

22. Marshall, “Grape iwi,” *Occasional: 50 Poems*, 52.

23. Marshall, “Relativity and the Solar System,” *Occasional: 50 Poems*, 53–54.

24. Bill Manhire, “Into the Ice Shelf,” *Collected Poems* (Wellington: Victoria University Press, 2001), lines 1–6.

25. Bill Manhire, *Doubtful Sounds: Essays and Interviews* (Wellington: Victoria University Press, 2000), 244.

26. Manhire, “Deep Field Song,” *Collected Poems*, lines 1–2, 6, 11–12.

27. Manhire, “Some Frames,” *Collected Poems*, lines 1–8.

28. Manhire, “A Poet at the Pole,” *Doubtful Sounds: Essays and Interviews*, 244.

29. Ibid.

30. Bill Manhire, “Herschel at the Cape,” *The Victims of Lightning* (Wellington: Victoria University Press, 2010).

31. Bill Manhire, “Erebus Voices,” *Lifted* (Wellington: Victoria University Press, 2005).

32. Manhire, “Dogs,” *Lifted*, 36.

33. Ibid., lines 61–62.

34. Nigel Brown, Margaret Elliot, Margaret Mahy, Bill Manhire, and Chris Orsman. *Southern Convergence: Antarctic Art* (Wellington: Pemmican Press, 1999), 14.

35. Chris Orsman, *South: An Antarctic Journey* (Wellington: Victoria University Press, 1996).

36. Chris Orsman, *The Lakes of Mars* (Auckland: Auckland University Press, 2008).

37. Orsman, “The Book of the Dead,” *The Lakes of Mars*, lines 37–39.

38. Ibid., lines 55–58.

39. Orsman, “The Lakes of Mars,” *The Lakes of Mars*, line 7.

40. Ibid., lines 13–14.

41. Ibid., lines 25–27.

42. Orsman, “What the Camera Missed,” *The Lakes of Mars*, lines 19–21.

43. Rupert Summerson and Ian D. Bishop, “The Impact of Human Activities on Wilderness and Aesthetic Values in Antarctica,” *Polar Research* 31 (2012).

44. John H. Roscoe, “Exploring Antarctica Vicariously: A Survey of Recent Literature,” *Geographical Review* 48, no. 3 (1958): 409.

45. Stephen Nicol, “Antarctica: Surround Sound,” in *Antarctica: Music, Sounds, and Cultural Connections*, eds. Bernadette Hince, Rupert Summerson, and Arnan Wiesel (Acton: ANU Press, 2015), 167.

46. Antarctica Treaty, “Resolution Two: Education and Cultural Activities” (Utrecht: Antarctica Treaty Consultative Meeting, 1996).

BIBLIOGRAPHY

Antarctica Treaty. "Resolution Two: Education and Cultural Activities" (Utrecht: Antarctica Treaty Consultative Meeting, 1996).

Brown, Nigel, Margaret Elliot, Margaret Mahy, Bill Manhire, and Chris Orsman. *Southern Convergence: Antarctic Art* (Wellington: Pemmican Press, 1999).

Glasberg, Elena. "Camera Artists in Antarctica," *New Zealand Journal of Photography* 65 (2007): 21–23.

Hall, Bernadette. "A Writer's Life, or, A Sackfull of Spuds," *The Lustre Jug* (Wellington: Victoria University Press, 2009).

———. "Rathcoola Rain," *The Lustre Jug* (Wellington: Victoria University Press, 2009).

———. "Door," *Life & Customs* (Wellington: Victoria University Press, 2013).

———. "How Ponting Thought He'd Died and Gone to Heaven," *Life & Customs* (Wellington: Victoria University Press, 2013).

———. "Le Monde Sauvage," *Life & Customs* (Wellington: Victoria University Press, 2013).

———. "The White Boudoir," *Life & Customs* (Wellington: Victoria University Press, 2013).

Leane, Elizabeth. "Isolation, Connectedness and the Uses of Text in Heroic-Era Antarctica: The Cases of Inexpressible and Elephant Islands," *Island Studies Journal* 2, no. 1 (2007): 67–76.

Juelskjær, Malou and Nete Schwennesen. "Intra-Active Entanglements – An Interview with Karen Barad," *Kvinder, Køn & Forskning* 12 (2012): 10–23.

Manhire, Bill. "A Poet at the Pole," *Doubtful Sounds: Essays and Interviews* (Wellington: Victoria University Press, 2000).

———. *Doubtful Sounds: Essays and Interviews* (Wellington: Victoria University Press, 2000).

———. "Deep Field Song," *Collected Poems* (Wellington: Victoria University Press, 2001).

———. "Into the Ice Shelf," *Collected Poems* (Wellington: Victoria University Press, 2001).

———. "Some Frames," *Collected Poems* (Wellington: Victoria University Press, 2001).

———. "Introduction," in *The Wide White Page: Writers Imagine Antarctica*, ed. Bill Manhire (Wellington: Victoria University Press, 2004).

———. "Dogs," *Lifted* (Wellington: Victoria University Press, 2005).

———. "Erebus Voices," *Lifted* (Wellington: Victoria University Press, 2005).

———. "Herschel at the Cape," *The Victims of Lightning* (Wellington: Victoria University Press, 2010).

Marshall, Owen. "At Pergamum," *Occasional: 50 Poems* (Christchurch: Hazard Press, 2004).

———. "Grape iwi," *Occasional: 50 Poems* (Christchurch: Hazard Press, 2004), 52.

———. "Marlborough," *Occasional: 50 Poems* (Christchurch: Hazard Press, 2004), 37.

———. "Relativity and the Solar System," *Occasional: 50 Poems* (Christchurch: Hazard Press, 2004).

———. “Sleepwalking in Antarctica,” *Sleepwalking in Antarctica and Other Poems* (Christchurch: Canterbury University Press, 2010).

———. “Snow Poem 3: Metaphysical Pole,” *Sleepwalking in Antarctica and Other Poems* (Christchurch: Canterbury University Press, 2010).

Orsman, Chris. *South: An Antarctic Journey* (Wellington: Victoria University Press, 1996).

———. “The Book of the Dead,” *The Lakes of Mars* (Auckland: Auckland University Press, 2008).

———. “The Lakes of Mars,” *The Lakes of Mars* (Auckland: Auckland University Press, 2008).

———. “What the Camera Missed,” *The Lakes of Mars* (Auckland: Auckland University Press, 2008).

———. *The Lakes of Mars* (Auckland: Auckland University Press, 2008).

Roscoe, John H. “Exploring Antarctica Vicariously: A Survey of Recent Literature,” *Geographical Review* 48, no. 3 (1958): 406–27.

Sewell, Bill. *Erebus: A Poem* (Christchurch: Hazard Press, 1999).

Stephen Nicol. “Antarctica: Surround Sound,” in *Antarctica: Music, Sounds, and Cultural Connections*, eds. Bernadette Hince, Rupert Summerson, and Arnan Wiesel (Acton: ANU Press, 2015).

Summerson, Rupert and Ian D. Bishop. “The Impact of Human Activities on Wilderness and Aesthetic Values in Antarctica,” *Polar Research* 31 (2012).

Chapter 11

Mediating the "Deep"

A Partial Genealogy of Media Work with Oceans and Seas

Gareth Stanton

Ivan Illich wrote "The water that we seek is the fluid that drenches the inner and outer spaces of the imagination."[1] One way of imagining seas and oceans is through film and cinema, but nothing is straightforward in these worlds as this quotation from Iain Sinclair describing the 1956 film version of Melville's *Moby-Dick* suggests:

> Olson dropped in on Huston at the Warner Brothers studios. Jack Warner was away, until October, on the French Riviera. No hiring or firing could be undertaken. Or so the tactful Huston intimated. They talked books and existential philosophy (bluffing, both of them). The mechanical version of the White Whale wouldn't float in the studio tank. "Kill that fucking fish," Jack said on his return.[2]

The director, John Houston, intercut real footage of hunters chasing sperm whale off Madeira into the film itself. We must hold such factors in mind when we attempt to trace what we could at a stretch call "genealogies of nature," a process which helps us judge the historicity of certain attitudes and debates in the contemporary world and allows us to give a careful assessment of competing thinking. Attitudes to particular locales have changed considerably over time and how we regard, for example, the shoreline – hostile region to be avoided in the nineteenth century to holiday paradise (aka "beach") in the twentieth – enables us to make insightful judgements about contemporary environmental debates. Stacy Alaimo, who has written with great insight on many of the themes discussed in this chapter, has proposed that we inhabit a site she calls "trans-corporeality," "a place where corporeal themes and environmental themes meet and mingle in productive ways,"[3] such is the modest ambition of this short chapter.

SETTING SAIL

Philip Hoare's study of the human relationship to whales as described in his book *Leviathan or, the whale* provides a guide to the changing view of these animals as it has shifted from "natural resource" for exploitation (the "conquest of nature") to cherished guardian of the deep and keeper of our conscience in relation to the natural world[4] (encapsulated in the lyrics of Country Joe and the Fish's *Save the Whales* song).[5] Although this shift is not necessarily universal and some nations and groups, for "cultural reasons," still wish to hunt whales,[6] we can, nonetheless, trace a fundamental change in attitude that, in this instance, is framed by a canonical text, Herman Melville's 1851 novel, *Moby-Dick.*[7] While that book traces one man's murderous obsession with a white whale which leads to disaster, we can also identify an obsessive nodal point in relations with the natural world. The whale has been good to "think with" to use Levi-Strauss's expression. Indeed, it remains so and the obsession has contemporary forms such as killer whale displays at *Sea World* and various such modern "non-places" or foci of the "tourist gaze."[8]

That the contemporary obsession is no less damaging than that of Melville's Captain Ahab is demonstrated by a contemporary environmentally themed documentary such as *Blackfish.*[9] Our loves can also lead to death in bewildering and unexpected ways. This chapter hopes to demonstrate that the world of the oceanarium is not separate from the industrial spoliation of the sea but rather linked through complex webs of sedimented representation and commodification, orchestrated by various forms of media in the broadest sense that inform our imaginings and desires. As Barry Commoner emphasized "everything is connected" and the objective here is to trace these connections and make them visible.[10] Water and the oceans are crucial for building any debates about environmentalism and environmental thought.[11] Humans have been dependent on both from the earliest times so they form highly complex sites for symbolic and cultural thought. Any attempt to unpick the current environmental debates and their many articulations around scarcity and pollution need to take the broad view and trace the development of the idea of the ocean and map the imaginative spaces that it occupies.

We now increasingly talk of the Anthropocene and the debate multiplies and metastasizes like some new academic Klondike. At such a time the thoughtful among us would do well to reflect that the vagaries of the long aeons we call geological time are really just a version of another sort of time – oceanic time. The multitudinous denizens of the deep have been laying down their lives for untold millennia and the discovery of their impacted and fossilized bodies gave rise eventually to the science of geology which itself has imparted the categorization system to which so much printer ink has been devoted of late. It is with this in mind that we can conceive of geology itself as, in some sense, a human process of visualizing the oceans themselves, those parts of our shared world which are often regarded as the most unknown in contemporary times. A representative text here might be Richard Fortey's, *Trilobite!: Eyewitness to Evolution* – "They watched through their crystal eyes whilst life evolved, their own evolution calibrating geological time itself," states the book's publicity blurb.[12] These processes of visualization are worth considering in some depth and, indeed, at some depth because the representative strata that the Anthropocene (should we bow to the usage) will bequeath will be thinner than a computer chip.

This chapter then will seek to outline what it means to see the ocean as we have entered the modern media age. I will characterize this in a slightly perverse way some might feel, but it means to me the beginnings of the period of what Elizabeth Kolbert calls the "sixth extinction."[13] This will seem deliberately vague to many, but it is situated at the cusp of what Marx called primitive accumulation or plunder and more recognizable forms of modern industrial capitalism – when plunder itself became a zero-sum game in respect of the natural world. The time when humans started to take out species in systematic and quasi-industrial way and endangering not just whales but all life in the ocean towards a time where there are increasingly no fish in the sea, the era of the "fishing-industrial complex."[14] This is perhaps best summed up by eco-criticism pioneer Laurence Buell who describes one of central themes of Melville's *Moby-Dick* – the opposition between whale men bestialized by their work of hunting and "whales maddened by being hunted" – as being "culturally avant-garde insofar as it implies a comparative pathology of early capitalist enterprise and of intelligent mammals under pressure of systematic harassment."[15]

BLUE CULTURAL STUDIES: A NEW THALASSOLOGY?

For my own part, I was always interested in fishes and, as soon as I could, I filled my life with them, lining the walls of my bedroom with small tanks that I made cheaply out of panes of glass recycled from the decrepit greenhouse that sat in the garden. It was in my bedroom, basking in the glow of that wall of tanks, that I received my first romantic kiss. The obsession declined once, as my father later said, I had discovered "birds, booze and fags"[16] – they would hold me in their grip for the next forty years, but I still recall the aquarist shops and the lonely men that ran them and the silent lonely boy. Living in Tunis in the early 1980s, a recipe for seahorse couscous emanating from the coastal town of Sfax perplexed me. When I think of the cultural formations of this period in the United Kingdom, a picture does emerge with the British, by and large, having fallen out of love with fish, with the exception of the "national dish," "fish and chips." I remember wandering through the *medina* in Hammamet in and shopkeepers calling out "fish and chips" to the English tourists. Tunisia is not an idle context as it was here that Alan Davidson was based when he hatched the idea for his classic *Mediterranean Seafood*.[17] His main inspiration on the cooking front at the time, as for so many, was Elizabeth David, herself the author of the paradigm shifting *A Book of Mediterranean Food*, the book that catapulted British food culture out of the inward-looking world of post-war rationing and whale meat on menus.[18] Davidson took David's food revolution and focused it on fish. Not for him parsimony and frugality, he draws the attention of his readers to the Roman writer Apicius who, when his fortune had dwindled beyond recognition and he was no longer able to finance his gastronomic whimsies, committed suicide.[19] If some of our imaginations were being reconfigured in respect of food and the sea in the post-war period, the humanities were more focused on land.

More recently, however, this has changed. "The millennium is bringing humanities scholarship back to the sea" trumpets Steven Mentz as he proposes a "blue cultural studies."[20] Ignoring other connotations of blue, I am happy to follow him in this endeavour or borrow the other term he adopts from historians revising some of the orthodoxies of new historicism and updating the Mediterranean world of Braudel to proclaim a "new thalassology."[21] Furthermore, he writes, "In this moment of ecocriticism and environmentalism, the mind-stretching vastness of the sea provides powerful food for thought."[22] Just as well, we might add, given the increasing lack of food for food. Literary critic, Dan Brayton, in his study *Shakespeare's Ocean*, is critical of ecocriticism's "chlorophilia" and focuses his own work on bluer issues.[23] When looking at the ocean Philip Steinberg, author of *The Social Construction of the Ocean*, similarly suggests "the overall aesthetic is one borrowed from representations of the land."[24] For Steinberg, "It is only through engaging with the ocean in all its material complexity that we can develop the fluid perspective that allows us to use the sea to look beyond the sea."[25] Elsewhere he has suggested that the materiality of the ocean is experienced only by a small and often marginal group of people, despite an increased fascination[26] (one highlighted by such mega-series as *Blue Planet*[27] or the popularity of Animal Planet's *Whale Wars*). In summary, "Increased extraction of the ocean's resources has been accompanied by decreased integration of the ocean's material nature into everyday lives."[28] How can we grasp this paradox? The literary critics have one sort of answer. "The crisis of the marine environment," Dan Brayton suggests in his analysis of Peter Matthiessen's experimental novel *Far Tortuga* (1975), "is in important ways a literary phenomenon … and if we are to engage with the unsustainability of the human exploitation of the oceans, we must turn to the literature of the sea."[29] This new drift in thinking finds happy expression in the mixed catch of items contained in the wonderful book *Aquatopia* which accompanied a major exhibition at Nottingham Contemporary and Tate St Ives in 2013.[30] Featured here are classics from literature and art as well as some of the best recent writing and image making dealing with the imagination of the deep ocean.

MELVILLE'S "WHALE ETHNOGRAPHY": MOBY-DICK

Staying with the literary theme, postcolonial ecocritic Elizabeth DeLoughrey has also written of the rise of ocean studies and the recent turn to the sea, but, she suggests, quoting new Atlantic studies pioneer William Boelhower, we must take care how we proceed because "the sea leaves no traces, and has no place names, towns or dwelling places: it cannot be possessed."[31] However, her book *Routes and Roots* points to the fact that the nineteenth-century American maritime novel coincided with the point at which the United States became a global naval power.[32] The greatest of these novels is Melville's *Moby-Dick*, what Buell calls Melville's "whale ethnography."[33] For Hoare himself the novel "surpasses all other books because it is utterly unlike any other."[34] D. H. Lawrence, an early British convert to the book, called it "one of the strangest and most wonderful books in the world."[35] In truth though, Melville's literary reputation languished for over four decades. His death passed largely unnoticed. The assent of the reputation of his work and *Moby-Dick* in particular mirrored the very decline of the whale and whaling itself. This was brought about in accelerated fashion in the case of the former by such innovations as exploding harpoons, and in the latter by the end of some of the industries sustained by whale products, especially in the fashion industry, along with the discovery of other forms of oil and the beginnings of what Timothy Mitchell has called the age of "carbon democracy."[36]

It is fair though to echo Hoare when he writes, "More than anyone has realised, perhaps, the modern world was built upon the whale."[37] Although we can add murderous expansion in the New World and slavery. It might come as no surprise then to hear whale conservation scientist Roger Payne state, "I see in the history of the *conquistadores* something very similar to the history of the whalers."[38] There is also something about the descriptive horrors on offer in accounts of the whalers. Nathaniel Philbrick, Nantucket chronicler and author of *In the Heart of the Sea* a book based on accounts of the real life voyage of the *Essex*, which were themselves the inspiration for Melville's own epic.[39] He gives us this description:

> When the lance finally found its mark, the whale would begin to choke on its own blood, its spout transformed into a fifteen- to twenty foot geyser of gore that

> prompted the mate to shout, "Chimney's a fire!" As the blood rained down on them, the men took up the oars and backed furiously away, then paused to watch as the whale went into what was known as its flurry. Beating the water with its tail, snapping at the air with its jaws – even as it regurgitated large chunks of fish and squid – the creature began to swim in an ever tightening circle. Then, just as abruptly as the attack had begun with the first thrust of the harpoon, it ended. The whole whale fell motionless and silent, a giant black corpse floating fin-up in a slick of its own blood and vomit.[40]

The violence is graphic and cruel, and increasingly there were people who were starting to recognize this very horror.

It was Roy Chapman Andrews who wrote in 1916, "It is deeply to be regretted that the wholesale slaughter of the whales will inevitably result in their early commercial extinction."[41] The title of his book, *Hunting with Gun and Camera*, marks the birth of a different scopic regime for the whale. Andrews, Gary Kroll remarks, "made a sport out of hunting whales with his Graflex."[42] All this hints at the start of a different way of capturing the beasts and the start of a new consciousness of the whale. Still it will take almost a century of Melville scholarship and activism before we arrive at the subtle invocations contained in poet Dan Beachy-Quick's *A Whaler's Dictionary*. He retrospectively implicates all writers in the slaughter of the whale in his chapter "Mincer: A Writing Parable":

> To write a writer needs paper and light. Paper, too, can be burned for light. Form is also fuel. The whalers are in the business of light. To make light they make oil. They burn the oil from the blanket-pieces of blubber they unscroll from the whale. The thinner the blubber is cut, the more oil extracted. The mincer is the man who cuts the blubber. The crew shouts to him as he does so, "Bible leaves! Bible leaves!" so he remains mindful of the thinness desired. To light pages he makes pages.[43]

In the interim whales and the cetaceans in general have become the object of global spectatorship in themed sea-parks, in films and on TV. More surprisingly still, given these inauspicious beginnings, since the mid-1970s, as Susan Davis has observed, "Whales and dolphins have been caught up in a tide of popular spiritualism and cross-species identification."[44] For others, Melville's text itself has taken on this role in David Dowling's *Chasing the White Whale*, his quasi-ethnographic account of the Melville reading marathons held annually at the New Bedford Whaling Museum, he insists that "if you read *Moby-Dick*, you will be better."[45] In an interview with Melville scholar Samuel Otter, Thomas Farber, himself an eloquent chronicler of the contemporary Pacific, reads aloud from his book *The Face of the Deep* where he quotes Roger Payne's description of how "the whale enters our consciousness."[46] This is a paragraph also singled out by Buell.[47] Payne pronounced that Melville's *Moby-Dick* "is the greatest novel, the most inordinate book ever written."[48] Payne writes, Farber quotes, that Melville

> knew just how and by what steps whales would enter our minds, and how once inside they would metastasize and diffuse throughout the whole engine of human ingenuity, mastering and predisposing it to their purpose ... that whales would reconstitute themselves, reintegrate at the point of origin of all the meridians of the imagination, its very pole, and there tie themselves forever into human consciousness by a kind of zenith knot ... that when this process had completed itself the whale as symbol would have become the whale as puppeteer – would start orchestrating and manipulating, and directing the connections people perceive between themselves and the beating heart of nature ... that whales can help humanity save itself – help us to make the transition from Save the Whales to Saved by the Whales.[49]

LOOKING AT FISH

It is not just whales and other cetaceans, however, that need saving. Making a fetish of underwater creatures, extracting the ocean from the ocean so to speak, has inspired amateur aquarists ever since the Victorian period and fish lovers themselves through their very love of fish threaten their existence.[50] As we are inured to the very horrors of factory farming and large-scale extractive processes such as industrial fishing, so we are capable of ignoring the destruction that even our love of fish as pets not food, is capable of wreaking.[51] Keeping fish it might be argued is a form of false integration with the sea and, indeed, the depredations of the global pet trade mirror in some respects the decline of the fisheries, taking in even species not on the menu so to speak. In Pixar's *Finding Nemo* (2003), the sharks claim that "fish are friends not food" but even this film gives us pause to think that even being a friend (or pet) is barely an improvement on being eaten – at best a form of prison where release is dependent on the aquarium world equivalent of the IRA dirty protests at the Maze Prison in Northern Ireland.

Another way, of course, is to go and watch *in situ*. Jean Painlevé paved the way in this regard. Eva Hayward has recently examined his work and, in particular, his film *The Love Life of the Octopus (Les Amours de la pieuvre)* (1965).[52] Painlevé's "science cinema" emphasized the inventions of nature over the artifice of cinema.[53] For decades he had been experimenting with capturing and filming the world underwater and exploring the difficulties that this entailed. "The octopus," writes Hayward, "is under the refracting gaze of the camera, the panoptic frame of the aquarium, and the magnified look given to the spectator."[54] While not surrealists themselves, Painlevé and his collaborator, Geneviève Hamon, were experimenting in underwater cinematography: "They constructed underwater cameras and aquarium staging that allowed them to look into the unfamiliar worlds of common but strange marine organisms."[55] They modified existing technologies to their task. The invention by Yves le Prieur of a new diving apparatus offered the opportunity for utopian underwater living Painlevé thought and he dreamed of creating a complete studio entirely underwater. Hayward criticizes Painlevé's "persistent surveillance of animal physiology" and notes the cruelty inherent in the whole project. In further support of Hayward's position we can cite Painlevé's 1935 essay "*Les Pieds dans l'eau.*"[56] Here he writes of the frustration of trying to capture fleeting moments of marine life on camera. Having attempted to preserve them on film, however, the filmmaker's final reward, he remarks, is to be able to eat his "stars" – cooked in a myriad of different ways.

In another effort to resurrect this pioneer, Marina McDougall compares him with fellow Frenchman, Jacques Cousteau.[57] When Cousteau developed his new aqualung device he opened up the undersea world to the real scrutiny of untethered divers for the very first time. While the denizens of the deep had been fished and brought to the surface for millennia, life in the oceans themselves remained in an important sense a mystery. It had been left to the imagination of early science fiction writers such as the Frenchman Jules Verne to propose versions of the submarine world, but Captain Nemo was a fictional guide. Cousteau not only penetrated this hidden world, but he did so with new forms of visualizing technologies that could capture the life under the surface. He arrived in time for the new TV age and the films he made became part of that medium's love affair with the natural world. Indeed, just as television sought inspiration in the wonders of nature so did Disney and the big film studios. The increasing alliance between nature and popular entertainment would leave an indelible mark on attitudes to the environment in the developed world – precisely at a time when its own actual interactions with raw nature became increasingly limited. Cousteau's first book, *The Silent World* (1953),[58] was written in English demonstrating an unpatriotic pursuit of the American market. Of course swimming at record depths with "the freedom of fish," as the book's subtitle had it, placed Cousteau's work in a direct lineage with a previous generation's most famous explorer of the depths, American zoologist William Beebe.

Beebe and his sponsor, Otis Barton, made several deep dives in their famous Bathysphere.[59] One such dive in 1932 was broadcast by NBC in the United States and even heard on the BBC in Britain via a simultaneous shortwave radio link. Beebe's adventures in his Bathysphere were the basis for his best-selling 1934 book *Half Mile Down*.[60] While other biologists may have scoffed at Beebe's descriptions of the view at depth out of the Bathysphere, the general public felt they were actually visiting the worlds described by the likes of Verne. Barton tried to cash in on the excitement himself, directing a fictional movie released in 1938 and called *Titans of the Deep*. Surviving cinema posters depict vulnerable divers with heavy helmets defending themselves against a medley of sharks and giant octopi. Rachel Carson can be invoked here. Before the publication of *Silent Spring* in 1962, she was already well known as a popularizer of the marine environment. Beebe himself had encouraged Carson to undertake some dives of her own and had been instrumental in obtaining the fellowship that enabled her to write *The Sea Around Us* which was published in 1951 to great success.[61] The book was also made into an Academy Award-winning film in 1952. The underlying text of the movie has a contemporary ring – mankind may face extinction when the polar ice caps melt and the earth becomes covered with water. The general public were most impressed with the fight scene between a shark and an octopus. For her part, Carson was deeply unimpressed by the whole adventure and never sold the film rights to her own work again.

The 1960s may have been woken up to environmental issues by Carson, but they also witnessed the birth of *Flipper*,[62] a movie about a boy who befriends and is himself saved by a dolphin (a theme that had worked well for dogs in the past). *Flipper* in turn spawned a popular TV series. The boy cetacean bonding mood is recaptured later in the 1990s with the first of the *Free Willy* (1993) films.[63] This initial sizable crack in the Sea World pool, therefore, occurs just twenty years before the arrival of Cowperthwaite's indictment of captive orcas in her documentary *Blackfish*.

Meanwhile, out on the open seas, a new renegade boy/whale/man is arising in the shape of Greenpeace *refusnik* and founder of the Sea Shepard Conservation Society, Paul Watson. This scourge of the whalers wherever they are to be found gets his own TV show, *Whale Wars*. Some of this was foreseen by sometime environmental journalist Charles Bowden, who recounts an early voyage to disrupt whalers:

> Chicken Bob says, "Films is what it's about. In the early campaigns people would come on board and object to the video crew getting in the way of the action. They'd seen film of earlier campaigns on television and thought what they'd seen was what it was about. They didn't understand that taping what they'd seen was what it was about."
>
> We are all up in the Captain's room as he rolls the film of an old campaign. The Captain in his chair stares at the Captain on the screen. He mumbles a play-by-play as the action unfolds. Chicken Bob sits attentively as the Captain watches the Captain.
>
> Chicken Bob says, "This is it. Solid McLuhan, media, media, the image feeding the image, whales as light as a strip of film."[64]

Thomas Farber describes a diving expedition to Cocos Island, where hammerhead sharks school in large numbers, in similar terms:

> Nearly each diver, emerging, before coming on board the Zodiac hands up a camera, most of them costing thousands of dollars. There's a darkroom on the ship, film developed overnight – and video of course can be seen immediately. Thus almost instant metacommunication transpires, a viewing of simulacra of what we saw to … see what we saw? At times the images themselves appear to be the goal.[65]

Both Bowden and Farber here give us clues to a process which leaves the sea as pure spectacle and facilitates our removal from the "real" echoing the insights of Steinberg. Our love of the mediated sea, however, blinds us to the material sea and leads us to ignore its plight.

NO MORE FISH IN THE SEA

"I have hated fish and feared the sea and everything connected with it since I was two years old," wrote horror fiction pioneer H. P. Lovecraft.[66] He would not have been concerned that it is becoming increasingly clear the fish will not be saved (just as the whales may not have been). Elizabeth Kolbert writing about the depleted fish stocks of the world's oceans makes this clear, "Is there any hope for our overfished oceans or are we entering the terminal phase termed the Myxocene by fish conservationist Daniel Pauley?" she asks.[67] Pauley coined the expression in his book on ecological cascades *Five Easy Pieces*, and he himself has become one of the go-to scientists for journalists and documentary makers investigating the end of fish.[68] He derives this new geological time period from the Greek *muxa* for slime. That is all that will be left when the brief anthropocene comes to its inevitable end. That and the ubiquitous jellies that have survived the five previous great extinctions.[69] This should not come as news to anyone because the haul of recent popular books on the theme is extensive. A representative sample would include Farley Mowat's *Sea of Slaughter*, Colin Woodard's *Ocean's End*, Richard Ellis's *The Empty Ocean*, Charles Clover's *End of the Line*, Calum Roberts' *The Unnatural History of the Sea*, Teras Grescoe's *Dead Seas* and Mark Kurlansky's *World Without Fish*.[70] In his classic work of literary criticism *The Sense of an Ending*, Frank Kermode suggests that paradigms of apocalypse underlie much of our thinking about the world.[71] He argued that we are wrong to think that our own crises are in some way pre-eminent. The oceans of the past harboured huge monsters and leviathans but our mediation of these fears through science and other means have meant that fear of the depths has been replaced by fear of their imminent emptiness ("aquacalypse" as Pauly has it[72]).

"Melville's ocean is an arena of sudden violence," writes Buell,[73] and the old salts who tell their tales of terror do so "in words of mirth," as Melville states in the chapter on the try-works, where the blubber is rendered down in hellish conditions. This vision of love and hatred working together – dialectical opposites in some sense – is the structuring device of the original *Flipper* movie – the son's love of the dolphin and the father's hatred of the same creature. Some recent documentaries mine this same seam: the filmmakers' love of fish and the terrible realities of their treatment at human hands. The horrors that usher in the modern world reverberate here. In Rob Stewart's *Sharkwater* several of the talking heads make the comparison between slavery and the despoliation of the seas and the treatment of sharks.[74] One such terror highlighted in Stewart's film is the collateral damage caused by much fishing – the fraught question of by-catch. The film argues that if this was made properly visible, then it would not possibly be allowed. This is an argument that resurfaces in Rupert Murray's *The End of the Line*,[75] a documentary based on Charles Clover's book of the same name. We might call it the "if this were hunting on land thought experiment," and if it was cheetahs, say, being brought to the surface dead and not sharks or dolphins, would it be allowed? Helen Hughes writing about *Sharkwater* notes that it is a struggle to visualize the environmental community.[76] Citing Laura Marks' interesting speculations on animal documentaries,[77] she ponders the way in which the rights and needs of animals can sometimes be said to supersede those of certain sorts of humans. Marks uses the term "petishism," which she derives from the writing of film scholar Anne Friedberg.[78]

Nonetheless, no matter what fancy psychoanalytic gloss we chose to give such forms of inappropriate animal identification, there is a problem. In discussing the suffering of the shark Hughes cites Georges Franju's 1949 film *Le Sang des Betes* about the slaughterhouses of Paris. The brutal treatment of animals in that film is presented by a surrealist sleight of hand as if it is the hidden secret of modern society (the script is by Painlevé, it should be noted). This is a trick also attempted by Fernando Solanas and Octavio Getino's *Hour of the Furnaces* (1968) that classic of *Third Cinema*. And yet, as Hughes points out, such hidden secrets are pointed out time and time again – going all the way back to the Edison Manufacturing Company filming of the electrocution of Topsy the elephant in front of a paying audience at a Coney Island amusement park back in 1903. And yet each time they appear to be something new.[79] Elspeth Probyn takes a different approach and evokes the "problem with fish."[80] "How do we embody care for the sea and its dependants?" she asks. In a final footnote, she discusses Val Plumwood's assertion that she feels the same outrage at mass killings of humans as those of seals or dolphins. Probyn herself says she feels conflicted. Nonetheless, these are all zones of terror to evoke the anthropologist Michael Taussig.[81] The exploding harpoon has now been replaced by electrocution, states Roger Payne, the man responsible for drawing the world's attention to whale song, in *A Life Among Whales*.[82]

The Cove exposes the dark secrets of violence of the little Japanese fishing village of Taiji where an annual slaughter of dolphins occurs.[83] To oppose the slaughter is a personal quest for Ric O'Barry who trained the dolphins used in the TV series *Flipper*.[84] As a result of O'Barry's efforts, Richard Ellis observes, "Everybody was falling in love with whales and dolphins," but O'Barry became convinced that keeping such creatures in captivity was a crime after the "suicide" of one of his charges.[85] In a Pauline conversion he has dedicated the rest of his life to dolphin welfare and become a sort of anti-Ahab.[86] "When I started out," states O'Barry in the film, "there were only three dolphinariums … now at Teiji captured dolphins suitable for display fetch up to $150,000." The rest are slaughtered for their mercury-laden meat and fed to school children, the film alleges.[87] *Sharkwater* takes as one topos Cocos Island and Farber's nightmares are coming to pass. Line fishing is killing sharks and "finning" is already well established. Taiwanese factory vessels are disgorging their catches into secret wharfs in Puntarenas – fins without *fin*.

Probyn contrasts global mega-chef Gordon Ramsay unfavourably with another British celebrity chef, Heston Blumenthal, who appeared in a campaign for British retailer Waitrose "cuddling" a cod.[88] However, it is Ramsay's programme *Shark Bait* which recuperates much of the narrative of Stewart's film and serves it up again in heightened form.[89] The same places: Cocos Island, Costa Rica, Taiwan. The same deathly horrors: a six foot hammerhead is slowly and graphically finned and then lowered twitching back into the ocean. Discarded and disregarded on a trolley at the port of Puntarenas, Ramsey finds a baby hammerhead corpse (it could be a cuddly toy). He's feeling fucking sick you can bet. Even the *Daily Mail* found that some of his liberal use of the "f" word might have been appropriate in this particular instance as the fearless chef confronts fin dealers and shady oceanic shenanigans. He takes the caterers of London's China Town to one side and gets them to remove shark fin soup from the menus. Sadly the *Mail* has also unearthed other images – Gordon big-game fishing with his own hammerhead trophy taken off the Florida coast and shipped home, but only the head, no fins. Ramsey's alleged hypocrisy sells more papers than the impending extinction of a creature which has been swimming the oceans since before the dinosaurs.

As Probyn remarks, these documentaries do at least try to mobilize interest about fish and the sea.[90] She gets goose bumps listening to Daniel

Pauly's lightly "French-accented voice" – "All the fish are gone. Where are they gone? We have eaten them." Pauly grew up black in Switzerland – his escape was to the ocean via work in parts of the world where fish are often the only sustenance. He conducted basic fieldwork among fishing communities in the Philippines and Ghana, the experience giving him first-hand insight into the problems faced by such peoples in a world of diminishing fish stocks increasingly plundered by industrial fishing methods. Stewart's film notes that the Galapagos government was under pressure from local fishermen to allow previously banned line fishing (their incomes from sea cucumber having dropped due to overexploitation). Allegedly, the fishermen had threatened the last of the Galapagos tortoises. That would complete a cycle started by the whale men who prized their meat. The tortoises stayed alive for up to a year on ship without food or water. The sailors of the *Essex* had collected 180 tortoises before heading off to their doom.[91] Charles Darwin may have tasted their flesh. The *Beagle* carried away some thirty tortoises.[92] As biologist E. O. Wilson observed, "Humanity … eats its way down the food chain."[93]

CONCLUSION

I have tried to examine here some dimensions of our relationship with the sea and while much I have discussed has related to mediated visions in the form of films and literature I am also invoking the real material world of the oceans and bringing the two visions into collision. The whale and changing attitudes towards these wonderful beasts may give us some hope that things can improve, but our increasing distance from the material reality of the oceans, despite all our simulated and mediated "proximities" allows the continued spoliation of the oceans to proceed apace. Put simply and avoiding any theoretical posturing we have taken too many fish out of the sea and soon there will be none left. Around the edges there may be some room for tinkering – fishing restrictions, quotas and complex international agreements can be developed further. Increasing the scope of marine reserves would be another positive move. Developing forms of aquaculture that do not involve feeding wild fish to farmed fish at unsustainable ratios would improve the situation a little. Developing feed not based on other fish protein at all would help a little more. Greater consumer awareness about questions of sustainability and less destructive forms of fishing, about what is under threat and what can be eaten, cannot be bad things. Still, it is hard to see that our love of eating fish can be transformed into the vehicle of their salvation. Hard to imagine, as does the admirable fish conservation activist and restaurateur Casson Trenor, that we can save the oceans "one bite at a time."[94]

NOTES

1. Ivan Illich, *H2O and the Water of Life* (London: Marion Boyars, 1986), 24.

2. Iain Sinclair, *American Smoke* (London: Penguin Books, 2013), 44.

3. Stacy Alaimo, "Trans-Corporeal Feminism and the Ethical Space of Nature," In Stacy Alaimo and Susan Hekman (eds.), *Material Feminisms* (Bloomington: Indiana University Press, 2009), 238, 237–64.

4. Philip Hoare, *Leviathan or, the Whale* (London: Fourth Estate, 2008).

5. For an inspired and media theory inflected version of this see John Durham Peters, *The Marvellous Clouds: Toward a Philosophy of Elemental Media* (Chicago: University of Chicago Press, 2015).

6. See Arne Kalland, *Unveiling the Whale: Discourses on Whales and Whaling* (Oxford: Berghahn Books, 2009).

7. Herman Melville, *Moby-Dick; or, The Whale* (New York: Harper and Brothers, 1851).

8. Themes explored to great effect by Susan Davis in her ground-breaking book *Spectacular Nature: Corporate Culture and the Sea World Experience* (Berkeley: University of California Press, 1997). Tema Milstein's more recent work has extended a similarly critical ethnographic eye to the domain of whale watching in the wild – a logical progression from the Sea World experience and more environmentally friendly it might be thought. See, for example, Tema Milstein, "When Whales 'Speak for Themselves': Communication as a Mediating Force in Wildlife Tourism," *Environmental Communication: A Journal of Nature and Culture* 2, no. 2 (2008): 173–92.

9. Gabriela Cowperthwaite, *Blackfish* (CNN Films, 2013).

10. Barry Commoner, *The Closing Circle: Nature, Man, and Technology* (New York: Alfred A. Knopf, 1971), 33.

11. Lawrence Buell, *Writing for an Endangered World: Literature, Culture, and Environment in the U.S. and Beyond* (Harvard: Harvard University Press, 2001), 199.

12. Richard Fortey, *Trilobite! Eyewitness to Evolution* (London: HarperCollins, 2000).

13. Elizabeth Kolbert, *The Sixth Extinction* (New York: Henry Holt, 2014).

14. Daniel Pauly, "Aquacalypse Now." *New Republic*, September 28, 2009, accessed February 2016, https://newrepublic.com/article/69712/aquacalypse-now.

15. Buell, *Writing*, 214.

16. For the uninitiated in such UK idioms: women, alcohol and cigarettes.

17. Alan Davidson, *Mediterranean Seafood* (Harmondsworth: Penguin, 1972).

18. Elizabeth David, *A Book of Mediterranean Food* (London: John Lehmann, 1950).

In 1951 the innovation of *whacon*, corned whale meat, was introduced to the UK: all the goodness of whale meat without the fishy smell. It wasn't that popular either.

19. Davidson, *Mediterranean*, 226.

20. Steven Mentz, "Toward a Blue Cultural Studies: The Sea, Maritime Culture, and Early Modern English Literature," *Literature Compass* 6, no. 5 (2009): 997–1013.

21. Mentz, "Toward," 999.

22. Ibid., 1008.

23. Dan Brayton, *Shakespeare's Ocean: An Ecocritical Exploration* (Charlottesville: University of Virginia Press, 2012).

24. Philip E. Steinberg, *The Social Construction of the Ocean* (Cambridge: Cambridge University Press, 2001); Philip E Steinberg, "Of Other Seas: Metaphors and Materialities in Maritime Regions," *Atlantic Studies* 10, no. 2 (2013): 159, 156–69.

25. Steinberg, "Of Other Seas," 165.

26. Philip E. Steinberg, "It's So Easy Being Green: Overuse, Underexposure, and the Marine Environmentalist Consensus," *Geography Compass* 2, no. 6 (2008): 2080–96.

27. See Sean Cubbit, *EcoMedia* (Amsterdam: Rodolpi, 2005).

28. Steinberg, "It's So Easy," 2086.

29. Dan Brayton, "Writ in Water: *Far Tortuga* and the Crisis of the Marine Environment," *MPLA* 127, no. 3 (2012): 566, 565–71; Matthiessen's entire oeuvre is worthy of far greater attention. See in particular Peter Matthiessen, *Men's Lives* (New York: Random House, 1986).

30. Alex Farquharson and Martin Clark, eds., *Aquatopia: The Imaginary of the Deep* (Nottingham: Nottingham Contemporary, 2014).

31. Elizabeth DeLoughrey, "Heavy Waters: Waste and Atlantic Modernity," *PMLA* 125, no. 3 (2010): 705, 703–12.

32. Elizabeth DeLoughry, *Roots and Routes: Navigating Caribbean and Pacific Island Literature* (Honolulu: University of Hawa'i Press, 2007).

33. Buell, *Writing*, 209.

34. Hoare, *Leviathan*, 36.

35. Ibid., 37.

36. Timothy Mitchell, *Carbon Democracy: Political Power in the Age of Oil* (London: Verso, 2011). Although this did not cause the end of whaling, see Richard Ellis, *Men and Whales* (New York: The Lyons Press, 1999, orig. 1991). Indeed, despite new ways to illuminate homes and the end of the whalebone corset, the twentieth century still saw three million whales killed by whalers from 46 countries. See Sylvia Earle, *The World is Blue* (Washington: National Geographic Society, 2010), 42.

37. Hoare, *Leviathan*, 38.

38. Roger Payne, *Among Whales* (New York: Delta, 1995), 336.

39. Itself the inspiration for the movie of the same name directed by Ron Howard (2015).

40. Nathaniel Philbrick, *In the Heart of the Sea: The True Story that Inspired Moby Dick* (London: Harper Perennial, 2007), 54.

41. Quoted in Ellis, *Men and Whales*, 237.

42. Gary Kroll, *America's Ocean Wilderness: A Cultural History of Twentieth-Century Exploration* (Lawrence: University Press of Kansas, 2008), 18.

43. Dan Beachy-Quick, *A Whaler's Dictionary* (Minneapolis: Milkweed Editions, 2008), 177.

44. Susan G. Davis, *Spectacular Nature*, 9.

45. David Dowling, *Chasing the White Whale: The Moby-Dick Marathon; or, What Melville Means Today* (Iowa City: University of Iowa Press, 2010), 33.

46. Samuel Otter, "Facing the Deep: An Interview with Thomas Farber," *Leviathan* 8, no. 2 (2006): 55–73.

47. Buell, *Writing*, 207.

48. Payne, *Among Whales*, 324.

49. Thomas Farber, *The Face of the Deep* (San Francisco: Mercury House, 1998), 166.

50. Which can take on heroically fetishistic tones. See, for example, Chew Han Tah's documentary *Swimming with Dragons* (2014) about an obsessive keeper of Arowana. Available at: http://www.aljazeera.com/programmes/witness/2014/05/swimming-with-dragons-201452912241318 6942.html. More chilling is Sion Soto's (2010) film *Tsumetai Nettaigyo (Cold Fish).*

51. The gustatory impulse when combined with the admiring gaze gives us the Live Reef Food Fish Trade (LRFFT), which is devastating the so-called "coral triangle" region of the Pacific. This trade in live reef fish destined for the aquariums of fancy restaurants largely in Hong Kong, but increasingly also in the rest of China, has much larger mark-ups than merely dead fish. The diner pays a premium to select his/her own fish. It is awkward to note that criticisms of this trade and that in shark fins can come close to a new virulent strain of "Orientalism."

52. Eva S. Hayward, "Enfolded Vision: Refracting the Love Life of the Octopus," *Octopus: A Visual Studies Journal* 1 (2005): 29–44.

53. See Andy Bellows and Marina McDougall, *Science is Fiction: The Films of Jean Painlevé* (San Francisco: Brico Press, 2000).

54. Hayward, "Enfolded Vision," 38–39.

55. Ibid., 40.

56. A translation can be found in Bellows and McDougall, *Science*, 130–39.

57. Marina McDougall, "Introduction: Hybrid Roots," In Bellows and McDougall (eds.), *Science*, xiv–xviii.

58. Jacques Cousteau, *The Silent World* (London: Hamish Hamilton, 1953).

59. See Brad Matsen, *Descent: The Heroic Discovery of the Abyss* (New York: Pantheon, 2005).

60. William Beebe, *Half Mile Down* (New York: Harcourt, Brace and Co, 1934).

61. Rachel Carson, *The Sea Around Us* (Oxford: Oxford University Press, 1951).

62. James Clark, *Flipper* (MGM, 1963).

63. Simon Wincer, *Free Willy* (Canal+, 1993). It is easy to mock such films but read Diane Freedman, "A Whale of a Different Colour Melville and the Movies: The Great White Whale and Free Willy," *ISLE* 4, no. 2 (1997): 87–95.

64. Charles Bowden, *Some of the Dead are Still Breathing* (New York: Houghton Mifflin Harcourt, 2009), 176–77.

65. Farber, *Face*, 171.

66. Quoted in Sinclair, *American Smoke*, 66.

67. Elizabeth Kolbert, "The Scales Fall: Is There Any Hope for Our Overfished Oceans?" *New Yorker*, August 2, 2010, 70–73.

68. Daniel Pauly, *Five Easy Pieces: How Fishing Impacts Marine Ecosystems* (Washington: Island Press, 2010).

69. On this theme see Lisa-Ann Gerswin, *Stung! On Jellyfish Blooms and the Future of the Oceans* (Chicago: University of Chicago Press, 2013).

70. Farley Mowat, *Sea of Slaughter* (New York: Bantam Press, 1986); Colin Woodward, *Ocean's End* (New York: Basic Books, 2000); Richard Ellis, *The Empty Ocean: Plundering the World's Marine Life* (Washington: Island Press, 2003); Charles Clover, *The End of the Line* (London: Ebury Press, 2005); Callum Roberts, *The Unnatural History of the Sea* (London: Gaia, 2007); Teras Grescoe, *Dead Seas; How the Fish on Our Plates is Killing Our Planet* (London: Pan McMillan, 2009); Mark Kurlansky, *World without Fish* (New York: Workman Publishing, 2011).

71. Frank Kermode, *The Sense of an Ending* (Oxford: Oxford University Press, 1967).

72. Daniel Pauly, "Aquacalypse Now," *New Republic*, September 28, 2009, https://newrepublic.com/article/69712/aquacalypse-now.

73. Buell, *Writing*, 210.

74. Rob Stewart, *Sharkwater* (SW Productions, 2006).

75. Rupert Murray, *The End of the Line* (Arcane Productions, 2009).

76. Helen Hughes, "Humans, Sharks and Shared Environment in Contemporary Eco-Doc," in John Blewitt (ed.), *The Media, Animal Conservation and Environmental Education* (Routledge: New York, 2013), 25–40.

77. See Laura Marks, *Touch: Sensory Theory and Multisensory Media* (Minneapolis: University of Minnesota Press, 2002).

78. As does Eva Hayward – although the original coinage is Kathleen Szasz, *Petishism: Pets and their People in the Western World* (New York: Holt, Reinhart and Winston, 1969).

79. Although the problem of euthanizing whales takes a different form in Barry Lopez, "A Presentation of Whales," *Harper's Magazine*, March 1980, http://harpers.org/archive/1980/03/a-presentation-of-whales/.

80. Elspeth Probyn, "The Cultural Politics of Fish and Humans: A More-Than-Habitus of Consumption," *Cultural Politics* 10, no. 3 (2014): 292, 287–99.

81. Michael Taussig, *Shamanism, Colonialism, and the Wildman* (Chicago: Chicago University Press, 1987).

82. Bill Haney, *A Life among Whales* (Uncommon Productions, 2005).

83. Louie Psihoyo, *The Cove* (Diamond Docs, 2009).

84. There were several *Flippers*, just as *Shamu* was a generic name given to successive orcas at San Diego's *Sea World* to hide any deaths from the public.

85. Ellis, *Men and Whales*, 435.

86. See Richard O'Barry, *Behind the Dolphin Smile* (Chapple Hill: Algonquin Books, 1988).

87. Psihoyo, *The Cove*

88. Elspeth Probyn, "Les sentiment oceaniques," *Poli Image* 4 (2012).

89. *Shark Bait* (Channel 4 [UK], 2011).

90. Probyn, "Cultural Politics," 289.

91. Philbrick, *In the Heart*, 74.

92. Roberts, *Unnatural*, 401.

93. Quoted by Earle, *World*, 31.

94. Casson Trenor, *Sustainable Sushi: A Guide to Saving the Oceans One Bite at a Time* (Berkeley: North Atlantic Books, 2008).

BIBLIOGRAPHY

Alaimo, Stacy. "Trans-Corporeal Feminism and the Ethical Space of Nature." In Stacy Alaimo and Susan Hekman (eds.), *Material Feminisms* (Bloomington: Indiana University Press, 2009).

Beachy-Quick, Dan. *A Whaler's Dictionary* (Minneapolis: Milkweed Editions, 2008).

Beebe, William. *Half Mile Down* (New York: Harcourt, Brace and Co, 1934).

Bellows, Andy and Marina McDougall, *Science is Fiction: The Films of Jean Painlevé* (San Francisco: Brico Press, 2000).

Bowden, Charles. *Some of the Dead are Still Breathing* (New York: Houghton Mifflin Harcourt, 2009).

Brayton, Dan. "Writ in Water: *Far Tortuga* and the Crisis of the Marine Environment." *MPLA* 127, no. 3 (2012): 565–71.

———. *Shakespeare's Ocean: An Ecocritical Exploration* (Charlottesville: University of Virginia Press, 2012).

Buell, Lawrence. *Writing for an Endangered World: Literature, Culture, and Environment in the U.S. and Beyond* (Harvard: Harvard University Press, 2001).

Carson, Rachel. *The Sea Around Us* (Oxford: Oxford University Press, 1951).

Clark, James. *Flipper* (MGM, 1963).

Clover, Charles. *The End of the Line* (London: Ebury Press, 2005).

Commoner, Barry. *The Closing Circle: Nature, Man, and Technology* (New York: Alfred A. Knopf, 1971).

Cousteau, Jacques. *The Silent World* (London: Hamish Hamilton, 1953).

Cowperthwaite, Gabriela. *Blackfish* (CNN Films, 2013).

Cubitt, Sean. *EcoMedia* (Amsterdam: Rodolpi, 2005).

David, Elizabeth. *A Book of Mediterranean Food* (London: John Lehmann, 1950).

Davidson, Alan. *Mediterranean Seafood* (Harmondsworth: Penguin, 1972).

Davis, Susan. *Spectacular Nature: Corporate Culture and the Sea World Experience* (Berkeley: University of California Press, 1997).

DeLoughrey, Elizabeth. *Roots and Routes: Navigating Caribbean and Pacific Island Literature* (Honolulu: University of Hawai'i Press, 2007).

———. "Heavy Waters: Waste and Atlantic Modernity." *PMLA* 125, no. 3 (2010): 703–12.

Dowling, David. *Chasing the White Whale: The Moby-Dick Marathon; or, What Melville Means Today* (Iowa City: University of Iowa Press, 2010).

Earle, Sylvia. *The World is Blue* (Washington: National Geographic Society, 2010).

Ellis, Richard. *Men and Whales* (New York: The Lyons Press, 1999, orig. 1991).

———. *The Empty Ocean: Plundering the World's Marine Life* (Washington: Island Press, 2003).

Farber, Thomas. *The Face of the Deep* (San Francisco: Mercury House, 1998).

Farquharson, Alex and Martin Clark, eds. *Aquatopia: The Imaginary of the Deep* (Nottingham: Nottingham Contemporary, 2014).

Fortey, Richard. *Trilobite! Eyewitness to Evolution* (London: HarperCollins, 2000).

Freedman, Diane. "A Whale of a Different Colour Melville and the Movies: The Great White Whale and Free Willy." *ISLE* 4, no. 2 (1997): 87–95.

Gerswin, Lisa-Ann. *Stung! On Jellyfish Blooms and the Future of the Oceans* (Chicago: University of Chicago Press, 2013).
Grescoe, Teras. *Dead Seas; How the Fish on Our Plates is Killing Our Planet* (London: Pan McMillan, 2009).
Haney, Bill. *A Life among Whales* (Uncommon Productions, 2005).
Hayward, Eva S. "Enfolded Vision: Refracting the Love Life of the Octopus." *Octopus: A Visual Studies Journal* 1 (2005): 29–44.
Hoare, Philip. *Leviathan or, the Whale* (London: Fourth Estate, 2008).
Hughes, Helen. "Humans, Sharks and Shared Environment in Contemporary Eco-Doc." In John Blewitt (ed.), *The Media, Animal Conservation and Environmental Education* (Routledge: New York, 2013), 25–40.
Illich, Ivan. *H2O and the Water of Life* (London: Marion Boyars, 1986).
Kalland, Arne. *Unveiling the Whale: Discourses on Whales and Whaling* (Oxford: Berghahn Books, 2009).
Kermode, Frank. *The Sense of an Ending* (Oxford: Oxford University Press, 1967).
Kolbert, Elizabeth. "The Scales Fall: Is There Any Hope for Our Overfished Oceans?" *New Yorker*, August 2, 2010, 70–73.
———. *The Sixth Extinction* (New York: Henry Holt, 2014).
Kroll, Gary. *America's Ocean Wilderness: A Cultural History of Twentieth-Century Exploration* (Lawrence: University Press of Kansas, 2008).
Kurlansky, Mark. *World without Fish* (New York: Workman Publishing, 2011).
Lopez, Barry. "A Presentation of Whales." *Harper's Magazine*, March 1980, http://harpers.org/archive/1980/03/a-presentation-of-whales/.
Marks, Laura. *Touch: Sensory Theory and Multisensory Media* (Minneapolis: University of Minnesota Press, 2002).
Matsen, Brad. *Descent: The Heroic Discovery of the Abyss* (New York: Pantheon, 2005).
Matthiessen, Peter. *Men's Lives* (New York: Random House, 1986).
McDougall, Marina. "Introduction: Hybrid Roots." In Bellows and McDougall (eds.), *Science*, xiv–xviii.
Melville, Herman. *Moby-Dick; or, The Whale* (New York: Harper and Brothers, 1851).
Mentz, Steven. "Toward a Blue Cultural Studies: The Sea, Maritime Culture, and Early Modern English Literature." *Literature Compass* 6, no. 5 (2009): 997–1013.
Milstein, Tema. "When Whales 'Speak for Themselves': Communication as a Mediating Force in Wildlife Tourism." *Environmental Communication: A Journal of Nature and Culture* 2, no. 2 (2008): 173–92.
Mitchell, Timothy. *Carbon Democracy: Political Power in the Age of Oil* (London: Verso, 2011).
Mowat, Farley. *Sea of Slaughter* (New York: Bantam Press, 1986).
Murray, Rupert. *The End of the Line* (Arcane Productions, 2009).
O'Barry, Richard. *Behind the Dolphin Smile* (Chapple Hill: Algonquin Books, 1988).
Otter, Samuel. "Facing the Deep: An Interview with Thomas Farber." *Leviathan* 8, no. 2 (2006): 55–73.
Pauly, Daniel. "Aquacalypse Now." *New Republic*, September 28, 2009, accessed February 2016, https://newrepublic.com/article/69712/aquacalypse-now.

———. *Five Easy Pieces: How Fishing Impacts Marine Ecosystems* (Washington: Island Press, 2010).

Payne, Roger. *Among Whales* (New York: Delta, 1995).

Peters, John Durham. *The Marvellous Clouds: Toward a Philosophy of Elemental Media* (Chicago: University of Chicago Press, 2015).

Philbrick, Nathaniel. *In the Heart of the Sea: The True Story that Inspired Moby Dick* (London: Harper Perennial, 2007).

Probyn, Elspeth. "Les sentiment oceaniques." *Poli Image* 4 (2012).

———. "The Cultural Politics of Fish and Humans: A More-Than-Habitus of Consumption." *Cultural Politics* 10, no. 3 (2014): 292, 287–99.

Psihoyo, Louie. *The Cove* (Diamond Docs, 2009).

Roberts, Callum. *The Unnatural History of the Sea* (London: Gaia, 2007).

Shark Bait (Channel 4 [UK], 2011).

Sinclair, Iain. *American Smoke* (London: Penguin Books, 2013).

Soto, Sion. *Tsumetai Nettaigyo (Cold Fish)* (2010).

Steinberg, Philip E. *The Social Construction of the Ocean* (Cambridge: Cambridge University Press, 2001).

———. "It's So Easy Being Green: Overuse, Underexposure, and the Marine Environmentalist Consensus." *Geography Compass* 2, no. 6 (2008): 2080–96.

———. "Of Other Seas: Metaphors and Materialities in Maritime Regions." *Atlantic Studies* 10, no. 2 (2013): 159, 156–69.

Stewart, Rob. *Sharkwater* (SW Productions, 2006).

Szasz, Kathleen. *Petishism: Pets and their People in the Western World* (New York: Holt, Reinhart and Winston, 1969).

Tah, Chew Han. *Swimming with Dragons*, 2014. Available at: http://www.aljazeera.com/programmes/witness/2014/05/swimming-with-dragons-20145291224131869 42.html.

Taussig, Michael. *Shamanism, Colonialism, and the Wildman* (Chicago: Chicago University Press, 1987).

Trenor, Casson. *Sustainable Sushi: A Guide to Saving the Oceans One Bite at a Time* (Berkeley: North Atlantic Books, 2008).

Wincer, Simon. *Free Willy* (Canal+, 1993).

Woodward, Colin. *Ocean's End* (New York: Basic Books, 2000).

Chapter 12

Exiled in the Bush

A History of Landscape Transformation in Post-European Settlement New South Wales

David Orchard and Peter Orchard

Botany Bay has been described as a "landscape of encounter"[1]: an encounter not only between Indigenous Australians and European colonizers but also between colonizers and the bush. To settlers accustomed to the orderly, pastoral landscapes of England, the untamed Australian landscape must have seemed alien and intimidating. In part, the history of the development of Australia is the history of this confrontation, between a European mindset and a sometimes harsh Australian reality. More generally, the history of development in Australia is the history of demographic and socioeconomic change. Attitudes towards the environment have changed as attitudes towards Australia and Australianness have changed. The bush has been viewed not as a set of ecosystems to be preserved, but as an object to be transformed to meet the changing needs of its inhabitants, or as the raw materials from which an economic machine could be constructed. The landing at Botany Bay began an enduring process of domestication.

We can usefully subdivide this history into four phases: the expansion phase consisted of roughly a century of unrestrained development; the acclimatization period belongs to the free settlers that followed, who attempted to transform the bush into an image of England; the era of economic conservation reflects an increasing awareness of the value of the native flora and fauna, while still interpreting this value in primarily economic terms, conserving only those elements of the natural world that could be shown to have definite value; finally, modernity is characterized by an ambivalent attitude towards nature: a struggle between the acknowledged need to conserve the bush and the desire to fully exploit its economic potential, synthesizing past attitudes into an uneasy new whole.

While there is considerable value in viewing the environmental history of Australia in this way, it should be noted that these phases are not clearly delineated but overlap. The periodization of history represents a way of seeing, not a natural phenomenon that exists unambiguously in the world. The usefulness of this approach lies in its ability to crystallize the driving attitudes of a given period and identify the motives governing historical interventions into the landscape. If a common theme in our complex interrelationship with the landscape is a desire to make it conform to human desires, this periodization allows us to see clearly exactly when in history these desires have changed and identify the causes of this change. Previous studies have applied a historical framework to the consideration of rural sustainability from demographic and social perspectives[2] or have utilized resilience theory (the core concepts of which are iterative cycles of growth, accumulation, reorganization and renewal) to examine past changes and future adaptation on a regional scale.[3] Understanding historical changes in our attitude to the environment is the purpose of this chapter. We intend to present a survey of Australia's environmental history with an aim to uncovering the social, political and economic drivers of this change. In a world now facing extraordinary environmental crises in the form of climate change and loss of biodiversity, understanding how attitudes to the environment are shaped could be crucial.

EXPANSION

Early reports from Botany Bay were not encouraging. Joseph Banks warned of a "soil so barren … [that it] could not be supposed to yield much to the support of man."[4] Eighteen years passed between exploration and the eventual settlement, and this settlement was driven more by political necessity than by the promise of economic reward: the rebellion of the American colonies left Britain's prisons overcrowded, and the new colony represented a viable solution.[5] While the instructions issued to the colony's first Governor, Arthur Phillip, do make mention of England's desire to profit from the settlement (the decree compels Governor Phillip to "transmit a report of the actual state and quality of the soil …, [and] the probable and most effectual means of improving and cultivating the same"[6]), the first motive of Australian agriculture was simply to ensure the self-sufficiency of this prison colony.

Even this relatively modest goal presented a formidable challenge. Skilled labour was scarce, equipment was in short supply and the soils were deficient in essential nutrients. With no way of knowing which crops would be most suitable for Australian conditions, Phillip had been obliged to select a variety of species from a variety of climates.[7] From England the Fleet brought cereals, including wheat, barley, oats and maize; in Rio de Janeiro they took on board coffee, cotton, bananas, oranges, lemons, guavas, tamarinds, ipecacuanha and prickly pears (the last as hosts for the valuable dye-producing cochineal insect, *Dactylopius coccus*); and from the Cape of Good Hope they brought figs, sugar cane, quinces, apples, pears and strawberries. Of course, many of these plants died en route, and many more failed to thrive in the alien geography of the new colony. Those that survived the voyage were planted in two plots on the eastern side of Sydney Cove, red gums and swamp mahoganies making way for Australia's first crops. The clearing of the first farm was supervized by Henry Edward Dodd, a convict and personal servant to Governor Phillip. So poorly trained and so lacking in skill were the convicts that Phillip considered Dodd the only man among the convicts "who understood farming."[8] By October of 1788, Phillip had some 16 acres under cultivation, mostly growing cereals.

The early colony was dependent on supply ships from England. Visiting the Port Jackson site, Watkin Tench described Australia as "a country destitute of natural resource,"[9] and he was not alone in thinking this way. Crop failures occurred as early as 1796, attributed to exhaustion of the soil, and, by late 1805, convicts were expected to make a two-day ration of bread and salt pork last a week.[10] Attention therefore turned to other industries, and the harvest of red cedars, seals and whales kept the colony afloat. This enterprise resulted in the near-extirpation of red cedars from New South Wales (NSW) and the slaughter of (by some estimates) several hundred thousand whales and as many as 100,000 seals a season.[11] These industries exhausted themselves when their prey became scarce and therefore uneconomical.

Increasing free settlement and emancipation of convicts consolidated agriculture and made the colony viable. By 1819, some 857 emancipist settlers owned land in NSW. On average each had 46 acres to work with and owned 45 sheep and 28 cattle. At the same time, the 213 free settlers, many of them experienced agriculturalists, held on average some 900 acres and in some cases upwards of one thousand sheep.[12] Of these 900 acres only a small portion, perhaps 60 acres, would be cleared. The absence of labour prevented clearing on a large scale; thus early agriculture remained focused on grazing, not cropping, and could coexist to some degree with the bush.

Wool became Australia's dominant industry by 1830,[13] and the continuation of this boom through the 1830s encouraged the expansion of the sheep industry; by 1843, roughly 3 million sheep were grazed in the Nineteen Counties of Australia.[14] The scarcity of agricultural labour was solved by the gold rushes of the 1850s and 1860s, which saw a trebling of the population, from 437,665 in 1851 to more than 1.5 million by the late 1860s.[15] With few major gold finds to sustain the enthusiasm, many immigrants turned to agriculture, either as landholders or as labourers. Clearing could therefore by carried out on a larger scale, but the problem of soil quality persisted and made intensive cropping impracticable. Agriculture remained focused on grazing, bolstered by Britain's increasing dependency on agricultural imports.[16]

Despite attempts on behalf of the British authorities to restrict settlement to the Nineteen Counties of NSW, an area forming a rough triangle between Wellington in the west, Moruya in the south and the Manning River in the north, the rapid increase in population and the desire for land forced settlement inland. By 1870, only Cameron Corner, the most remote and arid area of NSW, remained unoccupied.[17] Occupation consisted of large, mostly uncleared properties inhabited by squatters. Soils were infertile and yields were poor – a problem solved not by improving agricultural techniques but by expanding the property boundaries to incorporate more land.[18] Attempts to break up the squatters' stranglehold began in the middle years of the nineteenth century, most notably with the passing of the Crown Lands Acts 1861 (NSW), which opened land up for conditional sale. Under these schemes, land ownership required improvement, meaning clearing and cultivation, to the value of £1 per acre.[19] Later schemes to encourage closer settlement resulted in the formation of properties that were too small to be economically viable without extensive clearing. Thus clearing became, unofficially at least, the policy of the colony, accelerating the degradation of the landscape.

In this way, it can be seen that the early history of post-European Australia was shaped by a political necessity which gave way to an economic motive. The philosophy that permitted unrestrained exploitation has its origins in many sources, beginning, of course, with the Bible: "And God blessed them, and God said unto them, Be fruitful, and multiply, and replenish the earth, and subdue it: and have dominion over the fish of the sea, and over the fowl of the air, and over every living thing that moveth upon the earth."[20] The triumph of man over untamed nature was also celebrated by Enlightenment thinkers, among them Johann Herder, who imagined mankind "as a band of bold though diminutive giants, gradually descending from the mountains, to subjugate the earth and change climates."[21] More generally, this thinking reflects the common acceptance of wealth as proof of industriousness and therefore virtue, a core component of the Protestant work ethic. Acquisition of resources becomes an essentially noble enterprise.

ACCLIMATIZATION

The dramatic increase in the Australian population resulting from the gold rushes did not immediately alter the direction Australian society was taking. Unassisted, often non-European migrants outnumbered assisted migrants during the gold rush years,[22] but it was the latter who occupied government and the professional classes. These were mostly English and the establishment remained essentially English. The colony had an increased sense of permanence by this time: no longer merely a penal colony, it was now a rapidly expanding and increasingly self-sufficient entity. The colony was not England, however, and the Englishmen in power knew it. The perceived deficiencies of the Australian landscape were to have a major impact on the direction the colony was to take.

To many, Australia was inhospitable to the point of being almost uninhabitable. In 1843, one British émigré offered the following lament: "At night no more he hears the delightful warblings of the queen of songsters – the charming nightingale. The hoarse croaking of the offensive bull-frog, and the incessant buzzing of the hideous mosquito, he takes in exchange for the gladdening tones of England's fairy songsters."[23] So beloved was the nightingale, in fact, that it was protected by law in NSW, even into the twentieth century, despite never having occurred in the colony.[24]

This Englishman, and those like him, were "exiled in the bush," a phrase coined by the zoologist Dr George Bennett in an 1862 lecture. Bennett counted himself among these exiles: so extreme was his homesickness that he found comfort in the cry of the catbird, which reminded him of the "nightly philharmonic concerts of the domestic cat" in London's streets.[25] (Bennett found other consolations in the bush, having eaten his way through the Australian fauna. Particular praise was lavished on the meat of bandicoots, wombats, possums and monitor lizards, and a soup made from kangaroo tails.[26]) Significantly, Bennett ended his lecture with a quote from Genesis, reminding his audience of man's dominion over nature but warning against destroying species through "mere wantonness."[27] What followed surpassed mere wantonness, however. Bennett, together with Charles Moore of the Sydney Botanic Gardens, was to found the NSW Acclimatisation Society, which in 1879 became the NSW Zoological Society and later the Royal Zoological Society of NSW. The stated purpose of the Society was not "to foster the study and conservation of Australia's unique native fauna," as it is now, but instead "the introduction and acclimatization of song birds and game."[28]

We can think of the process of acclimatization as an act of terraforming, or as the correcting or beautifying of Creation, and the middle years of the nineteenth century were a fertile time for this kind of thinking: nostalgic expatriate Englishmen, whether affiliated with an acclimatization society or not, were responsible for the introduction of the house sparrow, the common starling, the common blackbird, the song thrush, the goldfinch and greenfinch, carp, goldfish, mute swans and many others. The plagues of rabbits and foxes that have at various times threatened Australian agriculture were a by-product of acclimatizationist thinking. By and large, species were selected for food, fur and medicine, or for sentimental reasons. Sometimes the justifications were less clear: boa constrictors, eland and monkeys were all candidates.[29] Bennett did strongly emphasize the economic motive, however: "From a soil producing only a few fruits barely edible," he wrote, "we now obtain by Acclimatisation a large supply of food, luxuries, and important articles of commerce, affording subsistence for a large population of thousands of human beings."[30]

It was not only avowed acclimatizationists who felt the deficiencies of the bush. In his 1859 work, *Australian Facts and Prospects*, the poet R. H. Horne quoted with evident approval an unnamed writer to the *Argus* who stated that "nothing can be uglier" than Australia's native trees and advised that they be replaced. The writer proposed, with Horne's endorsement, that the new Australians "combine the vivid verdure of England with the luxurious vegetation of southern latitudes, and a land as fair as Italy would bask beneath a sky as blue, and a climate more benign."[31] There are clear echoes in these statements of the attitudes of Enlightenment thinkers expressed a century earlier, among them Buffon, who celebrated "cultivated Nature," "brilliantly and pompously subdued" by the efforts of man.[32]

The gazettal, in 1879, of Australia's first national park (then The National Park, now Royal National Park) appears at first glance to reflect a significant shift in focus towards conservation rather than exploitation. The motivation for this apparent act of beneficence was not conservation, however, but recreation.[33] Among the first National Parks to be declared worldwide, Royal was then 7,200 hectares of undisturbed vegetation. This was not to last: mangroves, woodlands and rainforests were cleared to make picnic lawns and ornamental gardens, to plant exotic trees, to dig quarries and to build dams and military installations. Areas were also set aside for the acclimatization of animal species. Deer, rabbits and foxes can all be found in the park today, descendants of those early introductions.

Acclimatization was short lived. The introduction of economically valuable species continued and continues, but not with such fervour. In time, economic arguments against acclimatization became more compelling than those for it, as the descendants of certain acclimatized species (including starlings, sparrows, rabbits and foxes among the animals, and Scotch thistle among the plants) began to wreak havoc on agriculture.[34] The sentimental argument lost ground too: generations were now growing up with no natural connection to England and its flora and fauna. An Australian nationalism was developing, which embraced the uniqueness of the continent's wildlife.

ECONOMIC CONSERVATION

The spread of European settlement and the introduction of exotic species was accompanied by the extinction of several native animal species and a decline in the abundance of many others. No attempt was made to preserve this fauna via legislation until the last decades of the nineteenth century, a period which also saw the mass extermination of animals deemed harmful to agriculture.[35] Calls for broad legislative protection of native animals had been made at least as early as the 1860s (e.g. by John Gould), but when protection came, beginning with the *Game Protection Act 1866* (NSW),[36] it was severely constrained in scope and only fitfully enforced. The first wildlife protection acts were invariably game acts, intended to establish close seasons and impose hunting restrictions. The *Native and Imported Game Act 1879* (NSW) is typical of this approach, levying fines for hunters who kill scheduled animals out of season or with prohibited firearms but stopping short of granting any animal outright protection. The philosophy behind this approach to conservation is outlined in the preamble to the Act: "It is expedient," it states, "to encourage the importation and breeding of Game not indigenous to the Colony of New South Wales and also to prevent the destruction of Native Game during the breeding season." In short, this Act, and those like it, were intended to protect animals for us, not from us.

Many acts of parliament followed in the subsequent decades, and the number of species under protection and the extent of the protection offered increased as federation approached. With the *Birds Protection Act 1901* (NSW), enacted in the year of federation, a small number of species were for the first time offered absolute, year-round protection. The species so protected, it should be noted, were all insectivores and included not only the nightingale, a bird never seen in Australia, but the common myna, now one of our worst animal pests. In the words of Alfred J. North, ornithologist and contributor to the list of protected species, the act was designed to preserve the "useful, edible and ornamental" species of Australia.[37] The focus on insectivorous birds reveals the economic motivation of these acts: species harmful to agriculture were generally excluded. For example, the inclusion of the king parrot (*Alisterus scapularis*) was protested by North, who thought protection should not be extended to that species "owing to its destructive habits in maize and other crops." This is a subtler form of terraforming than that espoused by the acclimatizationists, but it is similar in principle: nature should be made to conform to human will.

The increasing embrace of a conservationist ethos was motivated in part by nationalism. White notes, for example, that the national emblems adopted at federation prominently displayed some of the more charismatic and distinctive of the native animals, the red kangaroo and the emu.[38] It was also motivated, and perhaps more strongly, by economics. The protection of insectivorous birds has a clear economic motive: the elimination of pest insects from agriculture. This point is made clearly by Launcelot Harrison, zoologist and Secretary of the Wild Life Protection Society of Australia, a newly founded conservation body, in 1909: "In theory, bird protection on the economic plan is simple. Your bird eats so much that is helpful to man, and so much that is harmful. You simply balance one against the other, and if the good predominates, you protect the bird; if the bad, you condemn it."[39] To make this assessment, Harrison continues, detailed study would be required:

> To make our omelette of economic protection it will be necessary to break the eggs of bird lives. This may seem barbarous talk for an advocate of protection, but we will never induce the people to protect unless we can show them that it is to their advantage so to do; and all the Acts in creation will not succeed unless they have the force of public opinion behind them. It will be kinder in the long run to make a definite research and have done with it, and then to show sound reason for protecting such species as prove worthy of protection, than to allow people to exercise their own judgment, too often founded upon hasty observation and prejudice, in the destruction of birds.[40]

This sentiment was echoed by other prominent voices in the conservation debate, among them the entomologist W. W. Froggatt, who argued that "unless the people themselves are awakened to the beauty and value of our wonderful bird and animal fauna, no Act, however good, can be of much use."[41] Arguably, this awakening remains a significant hurdle to conservation and sustainability since political (policy-based) and economic drivers of landscape change operate under a social licence granted by a polity that seems largely uninterested. This theme is further developed in a later section that looks at creating change by means of a broader approach than the current biophysical focus allows.

Under the economic model of animal protection, for a bird to be worthy of our concern, there must be evidence of positive good and this evidence can only be obtained by rigorous study. With this in mind, from 1909 to 1916, the microbiologist J. Burton Cleland dissected specimens of the native bird life, hoping to identify their diet and thereby to determine the exact extent of their usefulness. Starlings and silvereyes, both omnivores, produced mixed report cards; the rainbow bee-eater was condemned as a pest to apiarists; and the butcherbirds, kingfishers, woodswallows, robins and many other small insectivores were all pronounced "useful."[42] In this same period (1910–1915), Froggatt published a lengthy series of articles in the *Agricultural Gazette of New South Wales* on the "Insectivorous Birds of New South Wales," communicating the research of Cleland and others to the general public. The same anthropocentric thinking that emphasized the usefulness or otherwise of animals found frequent expression in this series. In one entry he describes insects as "the most powerful group of enemies which the farmer and fruit-grower have to face."[43] He continues: "We owe a good deal of our protection to the Wood Swallows, which go out into the dry country and live upon the swarms, like so many little soldiers sent out by advancing agriculture to clear the ground of the enemy."[44] This is as clear an articulation of the notion that nature exists only for the benefit of human industry as one can find, and was echoed on many occasions by many authorities, among them J. R. Kinghorn, zoologist at the Australian Museum, who thought of birds as "the greatest army nature can muster" against the insect menace. Birds, he writes, "are man's most valuable but least valued possession."[45]

No protection was extended to mammals or reptiles until 1903 and when protection did arrive it was essentially arbitrary. In the second schedule of the *Birds and Animals Protection Act 1918* (NSW), 18 native mammals were denied protection. This includes some obvious selections, such as the dingo and the flying fox, but also some that are more surprising: the eastern quoll and bilby (both now presumed extinct in New South Wales), the eastern bettong (extinct on the mainland), the woylie (critically endangered) and the eastern hare-wallaby, a species not seen since 1889[46] or 1890[47]: almost thirty years after its apparent extinction, it could still be hunted. Curiously, the kangaroos, possibly the most common and most destructive to agriculture of the large grazing mammals, were protected. Perhaps by now they had become iconically Australian, and nationalist sentiment saved them. The rationale behind these selections is hard to fathom; the decision to protect certain species and not others seems to be based largely on anecdote and prejudice held over from generations of unrestrained hunting. Even as the law crept incrementally towards blanket protection, there was still the hope that uneconomic elements could be engineered out of the landscape.

In dramatically expanding the list of species granted protection, the *Birds and Animals Protection Act 1918* (NSW) can be regarded as a significant landmark. Rather than granting protection solely to species of known use, the act withheld protection from those species thought (rightly or wrongly) to do harm, thus reversing the *proof of good* ideology. Despite this apparent change of heart, this period (1890–1921) still saw the most rapid clearance of native vegetation in the history of NSW,[48] due partly to renewed efforts at encouraging closer settlement and partly to an agricultural boom brought about by the development of wheat varieties suited to Australian conditions, principally William Farrer's "Federation" variety. Large-scale clearing for wheat (and later other cereals) became more viable and as a consequence the flora suffered. While the later wildlife acts suggest an increasing awareness of the value of conservation, this understanding seems to have had little impact on the actual practice of individuals and organizations: the economic motive remained primary. Tellingly, though local governments could ban the collection of native plants from public lands, substantive legislative protection was not extended to native plants until the *Wild Flowers and Native Plants Protection Act 1927* (NSW),[49] and then only to horticulturally valuable groups such as *Boronia*, *Grevillea*, *Callistemon*, and *Dendrobium*, as well as ornamental palms and ferns.[50]

Thus, as we can see, conservation was motivated primarily by economic factors and secondarily by an increasing sense of Australianness. Prominent conservationists, including Froggatt and Harrison, tended to be Australian-born, unlike the acclimatizationists. Unlike Bennett, Froggatt and Harrison were not exiles in the bush: they spoke of the Australian fauna as "our" fauna, "our" birds. The increased prominence of figures like Froggatt and Harrison reflected the emergence and development of a professional scientific class, which also functioned to promote conservation. The professional interest in nature held by these eminent figures coexisted with a personal one. Both Froggatt and Harrison were actively engaged in environmental education, helping to promote an appreciation of the value of the bush. However, this scientific class would also encourage the modernization and development of the country, with serious consequences for the landscape.

MODERNITY

The modernization of agriculture was driven by mechanization, the development of agricultural chemicals, and the breeding of higher-yielding crop varieties. Thus the starting date of the modern era in Australian agriculture could be placed as early as the late nineteenth century, when the first agricultural institutes were founded, or as late as the middle years of the twentieth century, when mechanization peaked and agricultural chemicals, beginning with DDT and 2,4-D, began to see use. Regardless of the date one chooses, two incompatible themes emerge from this period: the intensification of agriculture, and the increasingly widespread application of conservation efforts. Significantly, this period saw rapid urbanization and a drift away from primary production towards white-collar employment.[51] Living in cities and no longer directly dependent on the land for their livelihood, the average Australian in this period had fewer direct interactions with the bush than their ancestors. The bush could become a site of recreation or a focus for personal or nationalistic sentiment, which in turn could foster support for conservation. At the same time, distance from the bush may encourage a diffusion of responsibility; lacking any personal engagement with the land, a newly urbanized Australian may no longer feel individually accountable for its protection. This in turn may feed into more exploitative attitudes towards the land's resources and encourage a focus on its potential economic value.

Modernization was an active process, driven by government, with the aim of maximizing the economic value of the landscape. Government-supported mechanization schemes caused the use of tractors and other farm machinery to increase rapidly. Between 1930 and 1950, the number of tractors in operation increased fourfold.[52] For larger holdings they were soon considered "essential,"[53] and the endorsement of the *Agricultural Gazette of New South Wales*, which urged farmers to "follow the pioneers,"[54] and the financial support of governments cemented this. Clearing and cultivation of the land became easier as a consequence: the intensification of agriculture came at the expense of the natural environment. Other pioneering technologies also came into use in this period. DDT and 2,4-D both saw use in the 1940s[55] and agricultural chemicals soon became ubiquitous, with profound ecological implications.

Superphosphate became the first chemical considered necessary for Australian agriculture,[56] as nutrient-poor Australian soils could not otherwise support the desired yields. Early experiments reported a strong response of clover-dominated exotic pastures to superphosphate, and while some native grasses also showed increased yields,[57] the performance of exotic grasses greatly exceeded that of the most common natives.[58] The pairing of clovers with superphosphate soon became standard, not only to improve the economic performance of existing pastures but to make uneconomic land profitable.[59] The strong response of exotic species to superphosphate favoured their widespread introduction. Lucerne expanded into the drier country and clover-based crop rotations proliferated. The sowing of clovers and exotic pasture species was also encouraged as a response to erosion, and by the middle of the twentieth century had become doctrine.[60] Intensive year-round production was no longer possible, it was reasoned, so maximum value had to be achieved in a smaller window of the year, and exotic species were the answer. In this way, the modernization of agriculture was the mechanization of the landscape: the transformation of its natural resources into an economic machine.

By the 1930s, the consequences of this unwise philosophy of modernization and land clearance were felt in widespread and destructive soil erosion. Mechanical fixes, such as contour banking, and changed agricultural practices (chiefly rotations) were widely encouraged, but tree planting was also proposed and would become enduringly popular. Agricultural publications, such as the *Agricultural Gazette of New South Wales*, began to advocate for the restoration of vegetative cover.[61] Though certain Australian trees, such as the kurrajong (*Brachychiton populneus*) became popular in revegetation,[62] the species selected for planting were not always native, and pines, poplars and willows now dominate some areas. Nevertheless, this represents the first significant attempt in Australian history to return vegetation to the landscape. Again, the motive was economic: the preservation of the landscape against a threat to agriculture. Early botanists (such as J. H. Maiden) had made some preliminary investigations of the native flora, and Ferdinand von Mueller had extolled the virtues of retaining eucalypts (for, among other things, preserving soil) half a century earlier[63]; only now, in a moment of economic necessity, was interest renewed.

Despite the intensification of agriculture and the consequent destruction of the landscape, the conservation imperative reached its widest acceptance in this period. Having long advocated for and provided methods (poisons, explosives) to aid in clearing the bush, the *Agricultural Gazette of New South Wales* began to encourage the restoration of the landscape. Here a writer for the *Gazette* urges farmers *not* to "follow the pioneers" in clearing the landscape:

> In their need to produce crops or ensure adequate pasture it was necessary for our pioneers, as an immediate task, to clear their lands of timber. ... However, their need for quick results and the apparently prodigal growth of scrub may excuse them, but not so those of their axe-happy descendants who seem to have accepted land clearing as a sacred tradition and are not content until they have slaughtered every tree in sight.[64]

What the first settlers did was necessary and forgivable, the author writes, but to their "axe-happy descendants" clearing should not be seen as a sacred right. This duality is the defining feature of modernity: owing to the advancement of technology we possess the ability to extract vast quantities of useful resources from the land, but more than ever before we know the consequences of this exploitation. This clash between short-term "exploitationism" and long-term conservationism provides the backdrop for all thinking on the environment today.

The history of post-European landscape transformation in Australia has been one of changing natural capital into economic capital, a process driven by government policies, operating under a social licence to achieve financial outcomes. Around 100 million hectares have been cleared largely for agriculture with over half a million hectares cleared in 2000.[65] An alternative calculation suggests that roughly 40 percent of forests have been lost but with an incalculable percentage of surviving forests degraded.[66] Although Australia possesses more than 500 national parks,[67] many of these are small, lack connectivity, and some, as former state forests with histories of active management, are severely degraded.

The change in landscape from largely perennial to significantly annual vegetation has resulted in leakage in water and nutrients, reduced ground cover and rising water tables, leading to soil salinity, erosion and acidification. Troublingly, these have generally been referred to and treated as problems rather than symptoms of land management practices and emphasis has been placed primarily on biophysical aspects and economic impacts. For example, the creation of the National Landcare Initiative in 1989 by federal and state governments with the provision of funds to local groups to implement natural resource management (NRM) programmes resulted mainly in tree-planting projects focused on soil salinity or erosion issues and as such had strong economic motives. An overview of these initiatives is given by Robins and Kanowski,[68] who were highly critical of *Caring for our Country*, a recent programme which conceptualizes environmental degradation as a form of market failure best addressed through market means, with success measured by short-term outputs.

Despite the efforts of Landcare and considerable expenditure on agricultural research by the Grains Research and Development Corporation (GRDC), a recent review of Australian cropping soils by Dang and Moody[69] noted that only 23 percent were without significant stresses. The impoverishment of the soil delayed and impeded the first European settlers, and these problems remain: salinity, sodicity, acidity, alkalinity, elemental toxicities such as boron, chloride and aluminium, and soil compaction were all listed by these authors, with some soils having more than one constraint. GRDC's External Investment Plan 2015–2016[70] also notes that lime use to counteract acidification was only 50 percent of that estimated to be needed, indicating that subsoils were vulnerable to increasing acidity. Responses to this included developing engineering solutions to apply lime at depth and breeding acid-tolerant plants, both of which continue the trend of addressing symptoms rather than problems. Further sustainability issues were highlighted in respect to chemical inputs to manage weeds, pests and diseases with increasing resistance in target species and a potential reduction in chemical options through regulation. It can be argued that despite a greater understanding of landscape processes, agricultural practices and the management techniques imposed on the landscape are still driven by economic outcomes: attitudinal changes have been essentially cosmetic. Further, even for many of those who broadly support conservation efforts, the bush remains remote. Kept at a distance, and without close attention to the science of sustainability, the degradation of the landscape may simply not register.

TODAY AND AFTER

Current landscape uses and practices are often justified on food security grounds: the urgent need to increase food production either through vertical increase in yield per unit of land or through horizontal increase in land available for agricultural production. Both of these strategies have sustainability and biodiversity implications. Recent interest has been shown in NSW in moving cropping eastwards into higher rainfall areas traditionally used for livestock production under perennial pasture systems. Annualization of the landscape will have marked impacts on hydrology, acidification, salinity and erosion with the potential to also introduce lower rainfall issues of herbicide resistance, weeds, pests and diseases into new areas. There is a bifurcation in our approach to NRM issues: while conservation is widely touted, the intensification of land use through increased mechanization and utilization of inputs such as fertilizers, pesticides and herbicides calls into question the sustainability of our agricultural practises.

Wilson argues that the guiding theme of Western civilization is the Exemptionalist view, wherein "our species exists apart from the natural world and holds dominion over it; we are exempt from the laws of ecology that bind other species; and there are few limits on our expansion."[71] There has never been a widespread ecological understanding or an acceptance of the limitations of the landscape. The short-term success of increasing yield in homogenous environments has tended to reinforce a belief in human development being independent of nature and allowed a focus on increasing economic efficiency and the control of natural variation to predominate.[72] There is a persistent belief that technological fixes will always be available and little recognition that frequently these are merely new Band-Aids, addressing symptoms and not the underlying problems. Alongside those who focus simply on producing better Band-Aids is a "business as usual" camp who tend to focus on resource use, economic growth and wealth generation and the use of market mechanisms for the efficient allocation of resources.[73] There is a tug of war between the demands on the landscape, and environmentalism is consistently the loser. Progress is incremental and frequently overturned: measures imposed by one government are reversed by the next, and enthusiasm for conservation waxes and wanes. The successes of the conservation movement have not altered the fact that landscape change has been consistently driven by political factors, under a social licence to achieve economic outcomes.

Hence, any action to reverse the loss of biodiversity and move towards sustainability needs to recognize that the problem is not simply biophysical (although some tools and solutions may be) and should incorporate disciplines other than science for wider societal engagement. As Williams and Saunders[74] and Harris[75] note, technology, data and information are only part of the story. New knowledge does not immediately translate to new policy or action: this depends on collaboration and communication between agencies, social and economic considerations, and even on individual values and beliefs. A new epistemology needs to be developed which may require the creation of new institutions[76] or a greater emphasis on transdisciplinary studies. Disciplines as diverse as demography, economics, ecology and epidemiology may be central to the sustainability argument, and other fields of study, including other social and natural sciences, engineering and the humanities, need to be engaged.[77]

Whether this can be achieved in a conventional university setting, with each discipline confined to its own silo, is an open question, though Kauffman[78] notes that academic institutions in the present century have increasingly tended to devote resources to multidisciplinary and cooperative research programmes. Sustainability science has evolved to become increasingly integrative and transdisciplinary, incorporating inputs from traditional disciplines – chiefly social, economic and engineering sciences – as well as local indigenous knowledge. Sustainability scientists, in consciously seeking to develop a more holistic discipline, work with a plurality of epistemologies, languages and styles of research, leading to a variety of epistemic and normative stances and methods. This collaborative approach may lead to novel methods of landscape management, in the manner of Tyler's Adaptive Co-management approach,[79] which emphasizes collaborative institutional arrangements and shared learning and tasks across a range of scales, driven by a common vision. Any scheme along these lines would face obstacles in the form of governance, given that legislative and sectoral mandates of agencies may be inflexible, subject to high degrees of political control, and even contradictory. Such institutions can be virtual rather than purpose built and populated on a voluntary basis[80] – that is, composed of members of pre-existing institutions working according to a common vision.

As long as environmental resources are seen by society primarily in financial and economic terms, however, little change will occur. Sustainable philosophies need to be embedded as a central core in all disciplines, as a matter of urgency. Moreover, if change requires both political will and social licence, issues of sustainability cannot remain confined to academia. The degradation of our environment is arguably the most urgent issue our species faces and the need for a genuine sustainability is pressing. There is little value in merging or bridging the traditional academic silos only to create a larger silo, from which the bulk of humanity is excluded. While the bridging of traditional disciplines offers opportunities to reach a wider audience, there is still a critical need to translate this increased awareness into actions and outcomes. The general awakening Harrison and Froggatt desired is still needed, and the sustainability ethos must be embraced and codified by policy makers. This cannot happen if the debate remains abstruse and academic. As Harris[81] suggests, the pursuit of sustainability also requires us to look at where we've come from and how we arrived at our present position, and to realize that the current state of our environment is not the norm but the cumulative result of our actions: generations of tinkering producing a profound effect over time.

NOTES

1. Doug Benson and Georgina Eldershaw, "Backdrop to Encounter: The 1770 Landscape of Botany Bay, the Plants Collected by Banks and Solander and Rehabilitation of Natural Vegetation at Kurnell," *Cunninghamia* 10 (2007): 113, https://

www.rbgsyd.nsw.gov.au/RoyalBotanicGarden/media/RBG/Science/Cunninghamia/Volume%2010%20-%202007/Cun101113Ben.pdf.

2. See Chris Cocklin and Jacqui Dibden (Eds.), *Sustainability and Change in Rural Australia* (Sydney: University of New South Wales Press, 2005).

3. Helen E. Allison and Richard J. Hobbs, "Resilience, Adaptive Capacity, and the 'Lock-in Trap' of the Western Australian Agricultural Region," *Ecology and Society* 9 (2004).

4. R. M. Younger, *Australia and the Australians: A Concise History*, 2nd Ed. (Richmond, VIC: Hutchinson Group, 1982), 49.

5. Ibid., 50–51.

6. Gwendolen H. Swinburne, *A Source Book of Australian History* (London: G. Bell and Sons Ltd, 1919), Part II, http://gutenberg.net.au/ebooks/e00094.html.

7. R. H. Anderson, "The First Farm," *Agricultural Gazette of New South Wales* 73 (November 1962): 561–62.

8. Anderson, "The First Farm," 588.

9. Watkin Tench and Tim Flannery, *1788* (Melbourne: Text Publishing, 2012), 92.

10. William J. Lines, *Taming the Great South Land: A History of Conquest of Nature in Australia* (Sydney: Allen & Unwin, 1992), 35.

11. Lines, *Taming the Great South Land*, 32–35.

12. A. G. L. Shaw, "History and Development of Australian Agriculture," in *Agriculture in the Australian Economy*, ed. D. B. Williams (Sydney: Sydney University Press, 1967), 2–3.

13. Shaw, "History and Development of Australian Agriculture," 3.

14. Michael Cannon, *Australia in the Victorian Age: Life in the Country, 3rd Edition* (South Yarra, VIC: Viking O'Neill, 1988), 16.

15. "Australian Historical Population Statistics, 2014," Australian Bureau of Statistics, accessed January 28, 2016, http://www.abs.gov.au/AUSSTATS/abs@.nsf/DetailsPage/3105.0.65.0012014?OpenDocument.

16. Cannon, *Australia in the Victorian Age*, 177.

17. Shaw, "History and Development of Australian Agriculture," 9 (Figure 1.1).

18. Cannon, *Australia in the Victorian Age*, 14–15.

19. "Archives in Brief 93 – Background to Conditional Purchase of Crown Land," The State Records Authority of New South Wales, accessed January 28, 2016, http://www.records.nsw.gov.au/state-archives/guides-and-finding-aids/archives-in-brief/archives-in-brief-93.

20. *Genesis* 1:28, King James Version.

21. Gregory A. Barton, "The Enlightenment Foundations of World Environmentalism," *Comparative Civilizations Review* 40 (1999): 8.

22. Shaw, "History and Development of Australian Agriculture," 5–8.

23. Walter W. Froggatt, "The Starling: A Study in Agricultural Zoology," *Agricultural Gazette of New South Wales* (July 2, 1912): 610.

24. Alfred J. North, "On Bird Protection and Bird Destruction in New South Wales," *Agricultural Gazette of New South Wales* (February 2, 1909): 141.

25. George Bennett, *Acclimatisation: Its Eminent Adaptation to Australia: A Lecture Delivered in Sydney* (Melbourne: Acclimatisation Society of Victoria, 1862), 22.

26. Peter Macinnis, "When Learned Men Schemed to Create a Land of Alien Species: The Acclimatisation Societies of the Late 19th Century," last modified February 24, 2009, http://members.ozemail.com.au/~macinnis/writing/alienspecies.htm.

27. Bennett, *Acclimatisation*, 40.

28. "RZS NSW History," Royal Zoological Society of New South Wales, accessed January 28, 2016, http://www.rzsnsw.org.au/#!history/c7yh.

29. Macinnis, "When Learned Men."

30. Ibid.

31. Richard H. Horne, *Australian Facts and Prospects* (London: Smith, Elder and Co., 1859), 186–87, https://books.google.com.au/books/about/Australian_Facts_and_Prospects.html?id=bQtFAAAAIAAJ.

32. Barton, "Enlightenment Foundations of World Environmentalism," 8–9.

33. R. B. Walker, "Fauna and Flora Protection in New South Wales, 1866-1948," *Journal of Australian Studies* 15 (1991): 25–26.

34. Macinnis, "When Learned Men."

35. Steven White, "British Colonialism, Australian Nationalism and the Law: Hierarchies of Wild Animal Protection," *Monash University Law Review* 39 (2013): 462–63.

36. Walker, "Fauna and Flora Protection," 17.

37. North, "On Bird Protection," 145.

38. White, "British Colonialism, Australian Nationalism and the Law," 468.

39. Launcelot Harrison, "Some Further Aspects of Bird Protection," *Agricultural Gazette of New South Wales* (October 2, 1909): 865.

40. Ibid.

41. Walter W. Froggatt, "Several Aspects of the Protection of Our Native Birds," *Agricultural Gazette of New South Wales* (May 3, 1909): 369.

42. See J. Burton Cleland, "Examination of Contents of Stomachs and Crops of Australian Birds," *Agricultural Gazette of New South Wales* (May 2, 1910) and J. Burton Cleland, "The Stomach Contents of Australian Birds," *Agricultural Gazette of New South Wales* (April 3, 1916).

43. Walter W. Froggatt, "Insectivorous Birds of New South Wales," *Agricultural Gazette of New South Wales* (April 2, 1910): 297.

44. Ibid., 234.

45. J. R. Kinghorn, "The Value of Birds to Man: Especially in Relation to Agriculture," *Agricultural Gazette of New South Wales* (January 1, 1924): 45.

46. "Eastern Hare-wallaby – profile," Office of Environment & Heritage, accessed January 28, 2016, http://www.environment.nsw.gov.au/threatenedspeciesapp/profile.aspx?id=20195.

47. Ian M. Johnstone, "Conservation of Species," *National Parks Journal* 43 (1999), accessed January 26, 2016, dazed.org/npa/npj/199906/.

48. Corey J. A. Bradshaw, "Little Left to Lose: Deforestation and Forest Degradation in Australia since European Colonization," *Journal of Plant Ecology* 5 (2012): 111.

49. R. B. Walker, "Fauna and Flora Protection in New South Wales, 1866-1948," *Journal of Australian Studies* 15 (1991): 23.

50. "Wild Flowers and Native Plants," *Agricultural Gazette of New South Wales* (October 1, 1944): 442.

51. "Australian Social Trends, 2000," Australian Bureau of Statistics, accessed February 17, 2016, http://www.abs.gov.au/AUSSTATS/abs@.nsf/7d12b0f6763c78caca257061001cc588/db7193812e1efc92ca2570ec000e215a!OpenDocument.

52. P. C. Druce, "Machinery Usage and Cost on Wheat Farms in the Central West," *Agricultural Gazette of New South Wales* (October 1, 1951): 512.

53. P. C. Druce, "The Business of Farming," *Agricultural Gazette of New South Wales* (November 1, 1948): 585.

54. L. Evans, "Follow the Pioneers," *Agricultural Gazette of New South Wales* (November 1, 1925): 822.

55. John Unsworth, "History of Pesticide Use," last modified May 10, 2010, http://agrochemicals.iupac.org/index.php?option=com_sobi2&sobi2Task=sobi2Details&catid=3&sobi2Id=31.

56. Terry Ryan, "The Australian Fertilizer Industry – Values and Issues: Australian Fertilizer Industry Conference 2010," accessed January 28, 2016, http://www.fertilizer.org.au/files/pdf/conf2010/speakers/Australian%20Fertilizer%20Industry%20Value%20and%20Issues%20-%20Ryan%20-%20Paper.pdf.

57. J. N. Whittet, "Pasture Improvement Experience in the Cowra District," *Agricultural Gazette of New South Wales* (September 1, 1926): 657–60.

58. For a summary of early experimental work, see F. H. Gruen, "Superphosphate Use in New South Wales," *Review of Marketing and Agricultural Economics* 23 (1955): 7–30.

59. H. C. Trumble, "Grassland Agronomy in Australia," in *Advances in Agronomy, Vol. 4.*, ed. A. G. Norman (New York: Academic Press Inc., 1952), 11.

60. See, for example, S. C. Hodgson, "Farming Systems and Soil Fertility: Factors in Soil Erosion Control," *Agricultural Gazette of New South Wales* (August 1, 1939); W. D. Hardy, "Soil Erosion Prevention and Control on the Southern Tableland," *Agricultural Gazette of New South Wales* (March 1, 1942); W. C. Duggan, "The Influence of Pastures on Soil Erosion Control in Eastern Riverina," *Agricultural Gazette of New South Wales* (October 1, 1941).

61. See, for example, "Plant Trees and Prevent Soil Erosion," *Agricultural Gazette of New South Wales* (May 1, 1934); E. S. Clayton, "Gully Erosion: A Serious Menace to Farming and Grazing Land," *Agricultural Gazette of New South Wales* (October 1, 1935); "Control of Wind Erosion: Re-establishment of Vegetative Covering," *Agricultural Gazette of New South Wales* (November 1, 1939).

62. See, for example, C. Walkden-Brown, "The Kurrajong: A Valuable Fodder and Shelter Tree," *Agricultural Gazette of New South Wales* (October 1, 1940).

63. Younger, *Australia and the Australians*, 368.

64. "Axes Were Made for Fools like Me," *Agricultural Gazette of New South Wales* (April, 1957): 169.

65. Jenny Goldie, "Population – the Great Multiplier," in *In Search of Sustainability*, ed. Jenny Goldie, Bob Douglas, and Bryan Furnass (Collingwood, VIC: CSIRO Publishing, 2005), 156.

66. Bradshaw, "Little left to lose," 109.

67. "National parks," Australian Government, last modified June 1, 2015, http://www.australia.gov.au/about-australia/australian-story/national-parks.

68. L. Robins and P. Kanowski, "'Crying for Our Country': Eight Ways in Which 'Caring for Our Country' Has Undermined Australia's Regional Model for Natural Resource Management," *Australian Journal of Environmental Management* 18 (2011): 88–108.

69. Yash Dang and Phil Moody, "Cost of Soil Induced Stress to Australian Grains Industry," last modified December 2013, http://www.grdc.com.au/~/media/F18281BE63B5493A8BC8F951AE92048B.pdf.

70. "GRDC External Investment Plan 2015–16," Grains Research & Development Corporation, accessed January 24, 2016, http://www.grdc.com.au/~/media/Documents/Applying-Reporting/External-Investment-Plan/GRDC-External-Investment-Plan-2015-16.pdf.

71. E. O. Wilson, *Consilience* (New York: Alfred A. Knopf Inc., 1998), 278.

72. Carl Folke et al., "Resilience and Sustainable Development: Building Adaptive Capacity in a World of Transformation," *Ambio* 31 (2002): 437–40.

73. Graham Harris, *Seeking Sustainability in an Age of Complexity* (Cambridge: Cambridge University Press, 2007), 224.

74. John Williams and Denis Saunders, "Land Use and Ecosystems," in *In Search of Sustainability*, ed. Jenny Goldie, Bob Douglas, and Bryan Furnass (Collingwood, VIC: CSIRO Publishing, 2005).

75. Harris, *Seeking Sustainability in an Age of Complexity.*

76. Ian Lowe, "Achieving a Sustainable Future," in *In Search of Sustainability*, ed. Jenny Goldie, Bob Douglas, and Bryan Furnass (Collingwood, VIC: CSIRO Publishing, 2005).

77. A. J. McMichael, C. D. Butler, and Carl Folke, "New Visions for Addressing Sustainability," *Science*, New Series, 302 (2003): 1919–20.

78. Joanne Kauffman, "Promoting Integration and Cooperation for Sustainability: Views from the Symposium Held at UNESCO Headquarters September 19, 2013," *Sustainability Science* 9 (2014).

79. Stephen R. Tyler, "Adaptive Learning in Natural Resource Management: Three Approaches to Research," last modified October 2008, https://idl-bnc.idrc.ca/dspace/bitstream/10625/46816/1/133266.pdf.

80. McMichael, Butler, and Folke, "New Visions for Addressing Sustainability."

81. Harris, *Seeking Sustainability in an Age of Complexity.*

BIBLIOGRAPHY

Agricultural Gazette of New South Wales. "Axes Were Made for Fools like Me" (April, 1957).

———. "Plant Trees and Prevent Soil Erosion" (May 1, 1934).

———. "Wild Flowers and Native Plants" (October 1, 1944).

Allison, Helen E. and Richard J. Hobbs. "Resilience, Adaptive Capacity, and the 'Lock-in Trap' of the Western Australian Agricultural Region." *Ecology and Society* 9 (2004).

Anderson, R. H. "The First Farm." *Agricultural Gazette of New South Wales* (November 1962): 561–62.

Australian Bureau of Statistics. "Australian Historical Population Statistics, 2014." Accessed January 28, 2016, http://www.abs.gov.au/AUSSTATS/abs@.nsf/DetailsPage/3105.0.65.0012014?OpenDocument.

Australian Bureau of Statistics. "Australian Social Trends, 2000." Accessed February 17, 2016, http://www.abs.gov.au/AUSSTATS/abs@.nsf/7d12b0f6763c78caca257061001cc588/db7193812e1efc92ca2570ec000e215a!OpenDocument.

Australian Government. "National parks." Last modified June 1, 2015, http://www.australia.gov.au/about-australia/australian-story/national-parks.

Barton, Gregory A. "The Enlightenment Foundations of World Environmentalism." *Comparative Civilizations Review* 40 (1999): 7–30.

Bennett, George. *Acclimatisation: Its Eminent Adaptation to Australia: A Lecture Delivered in Sydney* (Melbourne: Acclimatisation Society of Victoria, 1862), 22.

Benson, Doug and Georgina Eldershaw. "Backdrop to Encounter: The 1770 Landscape of Botany Bay, the Plants Collected by Banks and Solander and Rehabilitation of Natural Vegetation at Kurnell." *Cunninghamia* 10 (2007), https://www.rbgsyd.nsw.gov.au/RoyalBotanicGarden/media/RBG/Science/Cunninghamia/Volume%2010%20-%202007/Cun101113Ben.pdf.

Bradshaw, Corey J. A. "Little Left to Lose: Deforestation and Forest Degradation in Australia since European Colonization." *Journal of Plant Ecology* 5 (2012): 109–20.

Cannon, Michael. *Australia in the Victorian Age: Life in the Country, 3rd Edition* (South Yarra, VIC: Viking O'Neill, 1988).

Clayton, E. S. "Gully Erosion: A Serious Menace to Farming and Grazing Land." *Agricultural Gazette of New South Wales* (October 1, 1935).

———. "Control of Wind Erosion: Re-establishment of Vegetative Covering." *Agricultural Gazette of New South Wales* (November 1, 1939).

Cleland, J. Burton. "Examination of Contents of Stomachs and Crops of Australian Birds." *Agricultural Gazette of New South Wales* (May 2, 1910).

———. "The Stomach Contents of Australian Birds." *Agricultural Gazette of New South Wales* (April 3, 1916).

Cocklin, Chris and Jacqui Dibden (Eds.). *Sustainability and Change in Rural Australia* (Sydney: University of New South Wales Press, 2005).

Dang, Yash and Phil Moody. "Cost of Soil Induced Stress to Australian Grains Industry." Last modified December 2013, http://www.grdc.com.au/~/media/F18281BE-63B5493A8BC8F951AE92048B.pdf.

Druce, P. C. "The Business of Farming." *Agricultural Gazette of New South Wales* (November 1, 1948).

———. "Machinery Usage and Cost on Wheat Farms in the Central West." *Agricultural Gazette of New South Wales* (October 1, 1951): 512.

Duggan, W. C. "The Influence of Pastures on Soil Erosion Control in Eastern Riverina." *Agricultural Gazette of New South Wales* (October 1, 1941).

Evans, L. "Follow the Pioneers." *Agricultural Gazette of New South Wales* (November 1, 1925).

Folke, Carl et al. "Resilience and Sustainable Development: Building Adaptive Capacity in a World of Transformation." *Ambio* 31 (2002): 437–40.

Froggatt, Walter W. "Several Aspects of the Protection of Our Native Birds." *Agricultural Gazette of New South Wales* (May 3, 1909): 369.

———. "Insectivorous Birds of New South Wales." *Agricultural Gazette of New South Wales* (April 2, 1910).

———. "The Starling: A Study in Agricultural Zoology." *Agricultural Gazette of New South Wales* (July 2, 1912): 610.

Genesis 1:28, King James Version.

Goldie, Jenny. "Population – the Great Multiplier." In *In Search of Sustainability*. Ed. Jenny Goldie, Bob Douglas, and Bryan Furnass (Collingwood, VIC: CSIRO Publishing, 2005).

Grains Research & Development Corporation. "GRDC External Investment Plan 2015–16." Accessed January 24, 2016, http://www.grdc.com.au/~/media/Documents/Applying-Reporting/External-Investment-Plan/GRDC-External-Investment-Plan-2015-16.pdf.

Gruen, F. H. "Superphosphate Use in New South Wales." *Review of Marketing and Agricultural Economics* 23 (1955): 7–30.

Hardy, W. D. "Soil Erosion Prevention and Control on the Southern Tableland." *Agricultural Gazette of New South Wales* (March 1, 1942).

Harris, Graham. *Seeking Sustainability in an Age of Complexity* (Cambridge: Cambridge University Press, 2007).

Harrison, Launcelot. "Some Further Aspects of Bird Protection." *Agricultural Gazette of New South Wales* (October 2, 1909).

Hodgson, S. C. "Farming Systems and Soil Fertility: Factors in Soil Erosion Control." *Agricultural Gazette of New South Wales* (August 1, 1939).

Horne, Richard H. *Australian Facts and Prospects* (London: Smith, Elder and Co., 1859), https://books.google.com.au/books/about/Australian_Facts_and_Prospects.html?id=bQtFAAAAIAAJ, 186–87.

Johnstone, Ian M. "Conservation of Species." *National Parks Journal* 43 (1999). Accessed January 26, 2016, dazed.org/npa/npj/199906/.

Kauffman, Joanne. "Promoting Integration and Cooperation for Sustainability: Views from the Symposium Held at UNESCO Headquarters September 19, 2013." *Sustainability Science* 9 (2014).

Kinghorn, J. R. "The Value of Birds to Man: Especially in Relation to Agriculture." *Agricultural Gazette of New South Wales* (January 1, 1924): 45.

Lines, William J. *Taming the Great South Land: A History of Conquest of Nature in Australia*. (Sydney: Allen & Unwin, 1992), 35.

Lowe, Ian. "Achieving a Sustainable Future." In *In Search of Sustainability*. Ed. Jenny Goldie, Bob Douglas, and Bryan Furnass (Collingwood, VIC: CSIRO Publishing, 2005).

MacInnis, Peter. "When Learned Men Schemed to Create a Land of Alien Species: The Acclimatisation Societies of the Late 19th Century." Last modified February 24, 2009, http://members.ozemail.com.au/~macinnis/writing/alienspecies.htm.

McMichael, A. J., C. D. Butler, and Carl Folke. "New Visions for Addressing Sustainability." *Science*, New Series, 302 (2003): 1919–20.

North, Alfred J. "On Bird Protection and Bird Destruction in New South Wales." *Agricultural Gazette of New South Wales* (February 2, 1909).

Office of Environment & Heritage. "Eastern Hare-Wallaby – Profile." Accessed January 28, 2016, http://www.environment.nsw.gov.au/threatenedspeciesapp/profile.aspx?id=20195.

Robins, L. and P. Kanowski. "'Crying for Our Country': Eight Ways in Which 'Caring for Our Country' Has Undermined Australia's Regional Model for Natural Resource Management." *Australian Journal of Environmental Management* 18 (2011): 88–108.

Royal Zoological Society of New South Wales. "RZS NSW History." Accessed January 28, 2016, http://www.rzsnsw.org.au/#!history/c7yh.

Ryan, Terry. "The Australian Fertilizer Industry – Values and Issues: Australian Fertilizer Industry Conference 2010." Accessed January 28, 2016, http://www.fertilizer.org.au/files/pdf/conf2010/speakers/Australian%20Fertilizer%20Industry%20Value%20and%20Issues%20-%20Ryan%20-%20Paper.pdf.

Shaw, A. G. L. "History and Development of Australian Agriculture." In *Agriculture in the Australian Economy*. Ed. D. B. Williams (Sydney: Sydney University Press, 1967), 2–3.

Swinburne, Gwendolen H. *A Source Book of Australian History* (London: G. Bell and Sons Ltd, 1919), Part II, http://gutenberg.net.au/ebooks/e00094.html.

Tench, Watkin and Tim Flannery. *1788* (Melbourne: Text Publishing, 2012).

The State Records Authority of New South Wales. "Archives in Brief 93 – Background to Conditional Purchase of Crown Land." Accessed January 28, 2016, http://www.records.nsw.gov.au/state-archives/guides-and-finding-aids/archives-in-brief/archives-in-brief-93.

Trumble, H. C. "Grassland Agronomy in Australia." In *Advances in Agronomy, Vol. 4*. Ed. A. G. Norman (New York: Academic Press Inc., 1952).

Tyler, Stephen R. "Adaptive Learning in Natural Resource Management: Three Approaches to Research." Last modified October 2008, https://idl-bnc.idrc.ca/dspace/bitstream/10625/46816/1/133266.pdf.

Unsworth, John. "History of Pesticide Use." Last modified May 10, 2010, http://agrochemicals.iupac.org/index.php?option=com_sobi2&sobi2Task=sobi2Details&catid=3&sobi2Id=31.

Walkden-Brown, C. "The Kurrajong: A Valuable Fodder and Shelter Tree." *Agricultural Gazette of New South Wales* (October 1, 1940).

Walker, R. B. "Fauna and Flora Protection in New South Wales, 1866-1948." *Journal of Australian Studies* 15 (1991): 17–28.

White, Steven. "British Colonialism, Australian Nationalism and the Law: Hierarchies of Wild Animal Protection." *Monash University Law Review* 39 (2013): 452–72.

Whittet, J. N. "Pasture Improvement Experience in the Cowra District." *Agricultural Gazette of New South Wales* (September 1, 1926): 657–60.

Williams, John and Denis Saunders. "Land Use and Ecosystems." In *In Search of Sustainability*. Ed. Jenny Goldie, Bob Douglas, and Bryan Furnass (Collingwood, VIC: CSIRO Publishing, 2005).

Wilson, E. O. *Consilience* (New York: Alfred A. Knopf Inc., 1998), 278.

Younger, R. M. *Australia and the Australians: A Concise History*, 2nd Ed. (Richmond, VIC: Hutchinson Group, 1982).

Chapter 13

Mapping the Anthropocene

Sy Taffel

The narratives we propagate about our past and present actions have important consequences for the future that comes to pass. The Anthropocene is one way of naming, defining and thereby shaping perceptions and responses to numerous contemporary ecological crises, which has rapidly gained popularity since Eugene Stoermer and Paul Crutzen introduced the term at the turn of the century.[1] Currently, the Anthropocene stands on the brink of being formally recognized as an epoch within the Geological Time Scale (GTS); recommending that the Anthropocene constitutes a new unit of geological time was delegated to the Anthropocene Working Group, a body formed in 2009, who presented their findings in August 2016 at the 35th International Geological Congress, recommending that the Anthropocene be classified as a new geological epoch beginning around 1950. This recommendation has yet to be formally ratified by the International Union of Geological Sciences, and a particular Global Stratotype Section and Point (GSSP) marker – the "golden spike" which forms the necessary stratigraphic indicator for a new geological epoch – still has to be identified before the GTS will officially include the Anthropocene, a process which is forecast to take several years.[2]

There is, however, significant ongoing debate within the geological sciences surrounding the chronologies and causalities of the Anthropocene. This chapter seeks to produce a map of these many coterminous versions of the Anthropocene, and to contrast these claims with how commentators from the humanities and social sciences have critically engaged with a discourse which emerges from scientific disciplines, but which has consequences that far exceed the scope of the sciences. My argument here is that defining, naming and mapping plays a pivotal role in how we comprehend phenomena, and that this understanding informs our capacity to subsequently address, act and mobilize in response to the key ecological challenges of the twenty-first century. Consequently, the way we formulate the discourse of the Anthropocene has important ramifications for how we undertake contemporary modes of ecological activism and politics designed to address the range of entangled ecological crises associated with the term. The chapter examines how the process of constructing discourse is never a purely human, linguistic and cognitive affair, drawing attention to the ways that the Anthropocene is shaped through the use of mathematical and simulational technologies which allow assemblages of humans, computers, ice cores, fossils, sediments, sensors and an array of technologies associated with measurement to make predictive statements about the future. The constitutive role of technologies in our understandings of the planet and our relationships with it are thus one productive way of foregrounding the underlying entanglement of ecological relations implicated in the Anthropocene.

Examining numerous scientific and cultural rhetorics positing a range of potential GSSP markers for the Anthropocene reveals that there is no singular and definitive point of origination, while additionally querying the homogenization of humans implicit within the nomenclature of the "Age of Man."[3] In doing so, my aim is not to delineate a singular reproduction of the reality of geological transformations occurring across the past 11,000 years, but to draw together multiple strands from the past, whose convergence in the present performatively gestures towards a range of potential futures. Mapping is employed here as a productive practice, drawing upon the works of philosophers Gilles Deleuze and Felix Guattari, for whom mapping involves an experimental encounter orientated towards affecting the future, rather than tracing a static underlying reality:

> What distinguishes the map from the tracing is that it is entirely oriented toward an experimentation in contact with the real. The map does not reproduce an unconscious closed in upon itself; it constructs the unconscious. … The map is open and connectable in all of its dimensions; it is detachable, reversible, susceptible to constant modification. It can be torn, reversed, adapted to any kind of mounting, reworked by an individual, group, or social formation. It can be drawn on a wall, conceived of as a work of art, constructed as a political action or as a meditation.[4]

Whereas tracing reproduces a hermetically sealed world, mapping deliberately foregrounds particular features while omitting others, thereby comprising a performative activity which actively engages with a dynamic and mutable world. Mapping thus produces lines of flight, bifurcations which take complex, dynamical systems from one meta-stable state to another. In the context of the Anthropocene, where dystopian futures featuring planetary mass extinction events and societal collapse are ubiquitous, this means reconsidering our relationships with other humans, organic and technical systems to contemplate how we might construct geopolitical counter-narratives where catastrophe is not inevitable and totalizing.

ONE OR MANY ANTHROPOCENES?

In classifying the Anthropocene within existing scientific discourses, one pressing question addresses how this new unit of geological time fits into the existing GTS, which is of importance in delineating the scale and scope of the departure from current geological conditions that are being posited. While in vernacular discourses, the terms "era," "period," "epoch" and "age" are used interchangeably to signify extended temporal durations, within the GTS they refer to a hierarchically nested set of relationally defined chronological periods, with eras containing periods, which house epochs, which consist of ages. The GTS currently places the biosphere within the "Holocene epoch" (a term which literally translates as "entirely recent"), the second epoch of the Quaternary period. Merely 11,000 years in duration, the Holocene is still only beginning by geological standards, whereby epochs are routinely measured in millions of years. Consequently, there are questions whether the addition of the Anthropocene requires reclassifying the Holocene as an age within the Pleistocene, whether the Anthropocene becomes an age within the Holocene, or whether an epochal boundary is a sufficient unit of change to denote the mass extinction event which is predicted to define the Anthropocene. Previous planetary extinction events such as the Ordovician/Silurian and Triassic/Jurassic form boundaries between periods, the scale which sits above epochs in the GTS. Indeed, two of the most severe extinction events in the geological record, the Permian/Triassic and the Cretaceous/Paleogene form the boundaries between the Palaeozoic/Mesozoic and Mesozoic/Cenozoic eras, placing these striations two levels above that of epochal change. Consequently, cultural commentators such as Donna Haraway have argued that situating the Anthropocene as an epochal, rather than periodic, boundary underplays the significance of the widely predicted alterations to global biotic systems.[5]

Of the potential dates for the genesis of the Anthropocene, the earliest suggestion is 11,000 years ago, when humans began early forms of anthropogenic ecosystem engineering, including the domestication of species of plants and animals in North and South America, Asia and Africa. According to proponents such as archaeologists Bruce Smith and Melinda Zeder this approach is productive because it "broadens the scope of inquiry regarding human modification of the earth's ecosystems to encompass the entirety of the long and complex history of how humans came to occupy central stage in shaping the future of our planet."[6] This proposal suggests that we do not need to formalize a new geological epoch, as the Holocene and Anthropocene are identified as the same period, although a change of nomenclature to Holocene/Anthropocene is proposed.

Similar early Anthropocene hypotheses are proposed by paleoclimatologist William Ruddiman who focuses upon quantifiable alterations to Earth's atmosphere in the early Holocene which delineate a planetary scope for anthropogenic ecosystem engineering.[7] These modifications include rises in atmospheric concentrations of carbon dioxide approximately 8000 years ago which is attributed to clearing forests and woodland for early human agricultural practices, and a rise in atmospheric methane concentrations approximately 5000 years ago which is ascribed to the large-scale adoption of irrigated rice farming, particularly in China.[8] Ruddiman argues that rising methane levels reversed a natural global cooling trend and prevented the glaciation of Northern North America, thereby perceptibly altering climatic conditions, and furthermore, that quantifiable alterations to atmospheric methane levels evident within ice cores are capable of providing a primary GSSP marker.[9] Questions have been raised regarding this account of the Anthropocene, in terms of whether the measured increases in greenhouse gases are outside the range of natural variability found within the Holocene, and whether these alterations prevented continental glaciation as claimed.[10] Members of the Anthropocene Working Group additionally contend that the Anthropocene boundary should not demarcate the first traces of human impact visible within geological records, but must evidence significant changes to the Earth System which take place on a global scale and affect geological durations,[11] an approach that favours more recent dates where there exists greater evidence for such impacts.

While these reservations are articulated within the scientific literature, further questions arise when considering the discursive ramifications of the early Anthropocene hypothesis. If human agricultural activity dating back thousands of years marks the onset of the Anthropocene, the connotation is that current techno-scientific and industrial practices are part of a teleological trajectory where alternative social, cultural and political practices hardly seem to matter as humanity was always destined to decisively alter the biosphere.[12] If the Anthropocene is to be a call to action – a suggestion that humans need to drastically alter their current collective actions to avoid catastrophic consequences – then concluding that this process began thousands of years ago reduces the significance of the rupture which the Anthropocene posits and consequently seems to inhibit rather than inspire drastic changes aimed at reorienting societies towards resilient, zero-waste, carbon-neutral futures.[13]

Among the modern dates potentially marking the onset of the Anthropocene, the earliest suggestion is 1610, which aligns with a 7–10 ppm reduction in atmospheric carbon dioxide evidenced within polar ice cores.[14] Following colonial conquest, the human population of the Americas fell from approximately 60 million in 1492 to 6 million by 1650, and the carbon dioxide reduction accompanying this is primarily attributed to the resultant large-scale cessation of agricultural practices and subsequent reforestation within the Americas. Alongside this decline in atmospheric carbon dioxide, which Lewis and Maslin advocated as a GSSP marker, there is auxiliary supporting evidence such as the increased trans-oceanic movement of crops and animals known as the Columbian exchange[15] and that colonization and the global trade which consequently proliferated is referred to by historians and social scientists as the onset of the "modern world-system."[16] The primary criticisms of this proposed date have been that the quantifiable fall in carbon dioxide lies within the range of non-anthropogenically influenced variability found within the Holocene,[17] and comprises a miniscule perturbation to the Earth System when contrasted with the current rise associated with industrial activity, whereby concentrations have risen from 280 ppm in 1880 to over 400 ppm in 2015, an increase 12–17 times larger than the fall in the sixteenth and seventeenth centuries.[18]

While European colonialism is one sociological marker of rupture associated with the historical formation of the modern world system, Crutzen and Stoermer originally proposed 1784 as the genesis of the Anthropocene, synchronizing with the invention of the steam engine by James Watt.[19] This symbolically aligns the Anthropocene with the industrial revolution, modernity and the anthropogenic liberation of vast quantities of fossilized energy from the Earth, foregrounding the links between the Anthropocene and anthropogenic climate change. This period has numerous potential GSSP markers, including rising concentrations of atmospheric greenhouse gases and global temperatures, or the deposition of fly ash from the combustion of coal. The large-scale liberation of regional, and later global, deposits of fossil fuels is a crucial symbolic marker for those who seek to connect the Anthropocene to contemporary deleterious ecological practices, but this has been queried in terms of forming a clearly visible epochal boundary, largely as a consequence of the protracted spatial and temporal diffusion of the industrial revolution.[20]

The most recent date posited for the genesis of the Anthropocene is circa-1950, what is commonly termed "the great acceleration." After the Second World War we see numerous ecological markers exhibiting rapid rates of change including: rates of species extinction, reductions in stratospheric ozone, rises in atmospheric concentrations of greenhouse gases, disruptions to the nitrogen and phosphorus cycles and increasing oceanic acidification. While these alterations to the Earth System begin prior to 1950, the argument underlying the great acceleration argument is that the increased rate of change occurring almost simultaneously across numerous areas forms a significant marker.[21] The two most likely GSSP markers found here are the presence of radionuclides in tree rings as a result of anthropogenic usage of nuclear weapons and the greatly increased deposition of non-biodegradable industrial chemicals, metals and plastics which will persist as technofossils.[22] The great acceleration has been suggested as the optimal epochal boundary by members of the Anthropocene Working Group in several publications and consequently this comprises the most probable candidate for the officially recognized genesis of the Anthropocene.[23]

What emerges throughout these debates occurring within geological discourses of the Anthropocene is a central question surrounding how we comprehend, define and quantify processes of genesis and rupture and consequently how we construct boundaries within a meshwork of complex ecological relations. This foregrounds what Foucault, following Nietzsche, described as the myth of origins when outlining his archaeological and genealogical methods in opposition to the type of teleological history advanced by certain scientific approaches to the Anthropocene:

> Genealogy does not pretend to go back in time to restore and unbroken continuity that operates beyond the dispersion of forgotten things; its duty is not to demonstrate that the past actively exists in the present, having imposed a predetermined form on all its vicissitudes. Genealogy does not resemble the evolution of a species and does not map the destiny of a people.[24]

We can always keep tracing causal pathways backwards, but they do not necessarily conjoin to formulate a singular pathway that inexorably points towards the present. Perhaps there is something to be gained from shifting away from geology and an emphasis upon stratification, depth and boundaries towards thinking in the terms posited by archaeological stratigraphy; in terms of surfaces, assemblages and political relations unfolding through time and space.[25] Approached from this perspective, we move away from thinking about measuring neatly divisible epochal boundaries, towards considering how such striations are always porous attempts to map dynamic processes.[26]

CALCULATING THE FUTURE

Debates regarding the genesis of the Anthropocene can partially be understood as a consequence of geology's insistence upon rigid boundaries and quantifiable markers. We should, however, consider that the Holocene was classified with a GSSP formed by deuterium levels measured in a specific ice core, providing a date 11,650 years before present (defined as 1950) +/– 99 years.[27] A similar margin of error for the Anthropocene conceivably covers the period from 1781 to 1979; geologically speaking the difference between 1800 and 1900 is too small to be significant using traditional techniques. The point about the accuracy of how we measure the Anthropocene asks us to contemplate how and why we construct geological maps of the planet's past, present and future. As the accuracy of techniques and technologies associated with measurement have advanced, geologists and paleoclimatologists have been able to reconstruct elements of the past with far greater precision, so that there can now be disagreement as to which decade or year is optimal for the genesis of the Anthropocene. As science and technology studies has often sought to remind us, scientific knowledge emerges through the interaction of tools and humans, with technical instruments playing crucial roles in determining the types of insights produced from these encounters.[28] While discourse is often nominally located within the realm of ideas and language and is articulated in opposition to materiality and the domain of nature, when we consider the formation of scientific discourses, concepts and ideas are produced through the hybrid activities of humans, nonhuman "natural" actors such as rocks, trees and ice cores and the nonhuman technological actors which collectively afford the production of geological, palaeontological and climatological knowledge.

There is thus a dialectical technopolitics of measurement, in which the activity of processing quantifiable data using computational technologies is central to our collective awareness of contemporary ecological crises (and thus to any attempts to engage communal action designed to alter the current trajectory towards a planetary mass extinction event), while also comprising pivotal components of the mode of global capitalism associated with these crises.[29] Here we see what philosopher Bernard Stiegler describes as the pharmacology of technics, whereby technologies inescapably form both poison and remedy.[30] In *The Automatic Society*, Stiegler argues:

> The Anthropocene era is that of industrial capitalism, an era in which *calculation prevails over every other criteria of decision-making*, and where algorithmic and mechanical becoming is concretized and materialized as logical automation and automatism, thereby constituting the advent of nihilism, as computational society becomes an automatic and remotely controlled society.[31]

One key point established here, is that the widespread dissemination of calculative information processing technologies has led to a culture in which increasing amounts of data (personal, social, geological or climatological) can be measured, analysed and added to scalable, open-ended computational databases, what is typically referred to as "Big Data."

This technocultural infrastructure forms a feedback loop in which understandings of value come to increasingly be associated with what can be measured, calculated and quantified. This transformational process entails that value becomes narrowly defined within the realm of calculability, so what cannot be counted, digitized and databased effectively becomes devalued. This is one strand which connects the calculability of Big Data to neoliberal capitalism, where the biopolitical drive towards commodification bestows a quantifiable monetary value upon a range of living and non-living entities; again what cannot be quantified is devalued and externalized. Technologies of quantification are not neutral or static, they reproduce a logic predicated upon calculability, entailing that what can be captured and counted becomes understood to be valuable, whereas forms less amenable to representation as binary code become sidelined. Assessing how Big Data transforms processes of valuation in this way, the sociologist of technology David Beer insightfully describes them as forming "productive measures" which do not merely trace an external world, but comprise performative actants within socio-technical ecologies, prescribing certain types of valuation and behaviour while inhibiting others.[32]

When discussing the technological infrastructures and methodologies used within the scientific studies which form the evidence for Anthropogenic Climate Change and other markers of the Anthropocene, we should recognize that alongside measuring activity in the past and present, evidence derives from computational simulations of potential future events. Simulation is a methodology utilized for investigating phenomena that are not empirically observable because they: occurred in the distant past, or have not yet occurred; model temporal or spatial scales too large to measure; or explore systems that are so complex that direct observations do not yield useful results. In each case: "The simulation is based on a model constructed by the researcher that is more observable than the target phenomenon itself."[33] Regarding the Anthropocene, simulations are used to extrapolate upon present activities in order to model probable futures, predicting the impacts of contemporary actions and the likely impacts of an array of potential future scenarios.

Climate models are tested by examining how their results correlate with observable changes in Earth's past, and comprise one of the key strands of evidence suggesting that without radical action the twenty-first century will see catastrophic climate change involving an average temperature rise of approximately 4 degrees Celsius,[34] almost certainly eradicating large-scale human civilizations alongside millions of species of terrestrial and aquatic life. Claims surrounding sea-level rise are similarly derived from modelling probable futures, with simulations suggesting that by 2100 global sea levels may rise by two metres,[35] delineating that even if greenhouse gas emissions are drastically curtailed, many of the world's most populous coastal cities and low-lying island nations will become uninhabitable in the near future. The underlying logic of these simulations, that of predictive calculability, is equally pivotal to both the predictive algorithms which underpin social media such as Google's PageRank and Facebook's EdgeRank and the high-frequency trading algorithms which comprise over half of all stock market trading.[36] Predictive calculability therefore simultaneously enables forms of knowledge surrounding the probable ecological devastation of contemporary actions and foregrounds the urgent need for radical action, while producing the regime of predictive corporate dataveillance and 24/7 hyperconsumerism which create the nightmarish future envisioned by climatic simulations.[37]

We should additionally acknowledge that technologies involved in processes of measurement are not somehow external to the world and are materially complicit with the process of ecosystem engineering that we understand as the Anthropocene. Immense amounts of greenhouse gases emissions result from manufacturing and powering the supercomputers required to run simulations of the global climate,[38] launching satellites into orbit to measure tropospheric temperatures and conducting expeditions to polar regions where core samples up to three kilometres in length are extracted from the earth, cut into sections, then transported long distances and stored at temperatures below –15 degrees Celsius. This is not to suggest that these actions do not have positive effects by advancing knowledge, raising awareness and leading to forms of collective action, but that we must situate these practices as material actions with ecological consequences, demarcating a further way that technologies of measurement influence the world in ways that exceed simply measuring or representing it. Again, we see that technology is not simply a neutral implement, but acts in a pharmacological manner. Building rockets, computer networks and ice core drills requires enormous movements of a multitude of minerals, metals, plastics and fossil fuels.[39] While the ecological and geological maps they generate foreground the destructive tendencies of the system which produces them, they can only emerge from a globalized system of trade and production.

ANTHROPOCENE OR CAPITALOCENE

Alongside the scientific literature which has tended to embrace the taxonomy of the Anthropocene while debating how to coalesce disparate strands of stratigraphic evidence into a single quantifiable date, there have been numerous alternative taxonomies advanced by cultural theorists, foregrounding particular conceptual issues inherent to scientific discourses of the Anthropocene. The point of their inclusion here, is not to challenge the fundamental taxonomy of the Anthropocene, but to highlight specific ways that problematic epistemological assumptions are pervasive within discourses surrounding the geological "Age of (hu)Man(kind)," especially the notion that the onset of the Anthropocene marks the global visibility of human action upon a passive and inert nature. While such claims exemplify a dualistic approach which separates humans from nature, typically accompanied by contentions that nature can either be tamed through human technological advances or is a fragile entity which requires conservation from human actions, this section explores strains of contemporary cultural criticism which situate humanity within entangled assemblages, emphasizing that humans never act alone, or as a homogenous chorus.

Taking up these issues, eco-Marxist cultural theorist Jason Moore advocates that the Anthropocene: "Does not challenge the naturalized inequalities, alienation, and violence inscribed in modernity's strategic relations of power and production. … The mosaic of human activity in the web of life is reduced to an abstract humanity as homogeneous acting unit."[40] Rather than addressing the extreme inequality under which the planet's eighty-five richest individuals own as much wealth as the poorest 3.5 billion,[41] the Anthropocene addresses a homogenized humanity where impoverished Indian subsistence farmers are as much to blame for ecological ruin as corporate CEOs. Consequently, Moore advocates adopting the term "Capitalocene" to draw attention towards the economic mode of production which exists alongside the rise of the ecological and geological situation identified as the Anthropocene. Doing so, Moore contends, explicitly asks us to confront the inequity and violence perpetrated by certain groups of privileged humans and nonhumans towards other ecological assemblages, and therefore lifts the ideological veil of universalizing bourgeois Western discourse which pervades rhetorics of the Anthropocene.

While global inequalities of wealth indicate one line which pierces the notional homogeneity of Man, similar arguments can be advanced regarding divisions predicated upon gender and race. Kate Raworth has suggested the Anthropocene could more accurately be designated "the Manthropocene," highlighting the gendered dimension of the patriarchal societies under which anthropogenic climatic impacts have developed, alongside the gender make-up of the Anthropocene Working Group, which initially had one woman in a group of twenty-nine, but has subsequently expanded to five in thirty-six, undoubtedly an improvement, but nonetheless still less than 15 percent of the total membership.[42] Similarly, numerous authors within this volume demonstrate ways that the legacy of the universalist logic of colonialism pervades discourses of the Anthropocene (see Otter, Dawson, Grieves this volume). Often indigenous peoples have played minimal or non-existent roles in deforestation, intensive farming and fishing, emissions of greenhouse gases, or the usage of nuclear weapons, plastics, heavy metals and other activities understood to comprise the quantifiable changes identified with the Anthropocene. Despite this negligible role in its formation, indigenous people are frequently among the humans most deleteriously impacted by the Anthropocene.

Prominent examples include the residents of low-lying islands such as Tuvalu or the Maldives which are expected to be rendered uninhabitable by humans due to rising sea levels[43]; Inuit communities whose traditional ways of living are being severely disrupted by the rapidly melting Arctic[44] and Aboriginal communities whose lands have been used for uranium mining and whose health has consequently been detrimentally affected by the radioactive tailings generated as a by-product.[45]

Moore's proposed focus upon Capitalism rather than humanity as the source of ecological and geological changes has the additional advantage of moving away from addressing how a single species impacts external environments, towards considering how systems work within nature, acting in purposeful ways to reorientate entangled ecosystems. From this perspective, the issue is not restricted to human agencies and actions, but how multispecies assemblages including particular groups of humans acting alongside cows, corn, cats, computers, cell phones and other entities are collectively conducting modes of ecosystem engineering which instigate temporal rhythms of ecological change whose speed exceeds the adaptive capacities of a multitude of other species. The state of an ecosystem is never the result of a single form of life unilaterally taking action, as intimated by scientific discourses of the Anthropocene which contest that "the human imprint on the global environment has now become so large and active that it rivals some of the great forces of Nature in its impact on the functioning of the Earth system."[46] Such propositions understandably foreground the prominence of human activity in contemporary ecological transformation, but do so by situating human influences outside of "Nature," exiting within a rivalrous relationship. This restatement of a nature/culture dualism is also commonly encountered within popular discourses of environmentalism which ask us to "save the planet." While there exists some degree of nuance within the terms formulated by influential environmental activist and 350.org founder Bill McKibben, whereby anthropogenic climate change demarcates the end of a Nature which exists unaffected by human activity, humans are nonetheless not just one species among many, allowing us to identify unilateral human action as the cause of contemporaneous ecological crises.[47] This approach stands in contradistinction to various insights from feminist, science and technology studies, new materialist and posthumanism scholars, who have in various ways problematized dualistic ontologies, and particularly the nature/culture dualism.

Donna Haraway's works surrounding cyborgs, companion species and humans are particularly pertinent here their exposition of symbiosis and mutualistic relationships within multispecies assemblages.[48] Haraway has recently advocated approaching current ecological crises through a series of overlapping terms, "Anthropocene," "Capitalocene" and "Cthulucene," the latter term foregrounding the entanglement of a "myriad intra-active entities in assemblages – including the more-than-human, other-than-human, inhuman, and human-as-humus."[49] Rather than an ontologically distinct human who impacts nature, Haraway explores multispecies becomings-together and the new modes of kinship that she contends are required to construct alternative narratives which escape that of ecocide within the Anthropocene. Haraway reminds us that, "from the start the greatest planetary terraformers (and reformers) of all have been and still are bacteria,"[50] that no species ever acts alone, and that the microbial ecologies which have enacted ecological engineering for over 4 billion years still play pivotal roles which are often poorly understood by humans, including enabling the ability of vertebrates to digest food. A pertinent example of ecological change driven by microbial life is the largest scale mass extinction event in the geological record, the Permian-Triassic, in which (alongside a terrestrial biotic crisis driven by volcanic activity in the Siberian traps) approximately 96 percent of marine life was eradicated, largely as a result of bacterial blooms across the oceans which created large quantities of hydrogen sulphide which is toxic to other marine and terrestrial life forms, and depleted the oceans of oxygen. Bacterial transformations of the biosphere are not, however, reflected in the taxonomy of the GTS. Thus we see why there is a strong case that discourses of the Anthropocene, and naming a unit of geological time after a single species (which just happens to be us), continues the tradition of overstating human exceptionalism and separation from nature.

With regard to the merit of these proposed taxonomical interventions, we should highlight these pertinent critiques of the Anthropocene, but following McKenzie Wark, it is also crucial to contextualize these criticisms by explicating how the term "Anthropocene" fosters a dialogue between human and geological histories, potentially forging transdisciplinary connections and conversations between the sciences and humanities.[51] Rather than refusing or replacing an admittedly problematic term, Wark advocates accepting its shortcomings as a way of beginning a productive discussion with its proponents in the both the natural sciences and mainstream media culture. As Joanna Zylinska argues in *Minimal Ethics for the Anthropocene*, adopting a critical stance towards the Anthropocene should not be predicated upon its rejection, but is designed to develop a more responsible and complex discourse in order to mobilize political responses to it.[52] As Wark concludes, "Anthropocene it is then. A bad name for bad times."[53]

Fostering a dialogue with scientific discourses of the Anthropocene is of political importance when we consider the long-standing abuse directed at prominent climate scientists whose publications have unveiled the manner by which economic-business-as-usual produces calamitous impacts for a broad range of ecological systems. Climatologists such as Ben Santer, Michael Mann and Phil Jones have received death threats, had materials stolen and published online as parts of a well-orchestrated public smear campaigns, been (unsuccessfully) taken to court, and endured workplace and parliamentary inquiries into their academic publications (which have time and again found them to be reputable and honest scientists).[54] We should recall that the scientific evidence underpinning the Anthropocene is largely what enables the possibility of orchestrating political actions designed to partially address this situation, and that humanities and social science scholars can have a crucial role in foregrounding the politics of industry-led attempts to discredit scientific evidence. Palaeontology and geology have little to say about public relation strategies surrounding greenwashing and frequently shy away from overtly engaging in the politics and ethics of taking collective actions designed to enact ecological change. Consequently, it is crucial to stand with our colleagues in the sciences who have consistently brought these issues to the public attention, rather than distancing ourselves via an outright rejection of their terminology. This does not require deferring from critical engagements with the problematic anthropocentricism and universalism inherent to discourses of the Anthropocene, but that we ought to situate such critique within a collegial discussion with the sciences, leveraging the manner by which the Anthropocene creates a channel for communication between the supposedly discrete domains of nature and culture.

CONCLUSION

One of the central tenets of postmodern cultural theory was the supposed death of metanarratives.[55] The notion of human progress which underpinned Modernist political philosophies presupposed that over time not only human technical proficiency, but morality, kindness and knowledge would gradually improve, so successive generations of humans would live in increasingly peaceful, just and wealthy societies. Such a vision remained plausible after the Second World War under the paradigm of Keynesian social democracy despite the spectre of the Cold War and threat of technologically enabled mutually assured destruction, as wages rose, welfare states were constructed and many humans benefitted from technological advances in agriculture, medicine, communications and transportation. This was socially reinforced by the gradual erosion of the biological racism of previous centuries, the extension of the vote to women, the legalization of homosexuality and increased levels of national self-determination. Since the 1980s, however, the stagnation of average wages combined with rising economic inequality,[56] levels of unemployment, precarity[57] and awareness of local and global ecological crises have thoroughly punctured notions of a teleological drive towards enlightenment. Indeed, it has become increasingly recognized that new technologies bring with them novel capacities for social and ecological devastation as well as betterment,[58] and that technological advances do not necessarily go hand in hand with social and ethical progress.[59]

The Anthropocene is a provocative concept here, as it can be grasped as both the apotheosis of this lineage of thought and the outright rejection of postmodern postulates surrounding the death of metanarratives. The rejection of unidirectional progress is partially derived from a critique of the reductionism present within a singular vector of movement which homogenizes human actions and subsumes them within a teleological historical narrative, however, it is also distilled from a growing scepticism towards enlightenment promise that the future will present better material conditions for humans than the past. The Anthropocene can thereby be grasped as an inversion of the metanarrative of progress, with dystopian premonitions of a planetary mass extinction event supplanting utopian visions of enlightenment and prosperity.

The dystopian metanarrative of the Anthropocene as the inevitable human destruction of Nature reveals why it is politically necessary to maintain a sense of multiplicity and agency – albeit the constrained and relational modes posited by ecological and posthuman accounts of distributed being and multispecies assemblages rather than the classical notion of autonomous human agency – when engaged in discussions surrounding the potential for organized activities within the contemporary ecological context. We should bear in mind here that the scientific discourse laid out by the Anthropocene Working Group is designed as a call to action, foregrounding the impacts that humans are having on ecological and geological systems is designed to focus attention upon these activities and lead to behavioural change:

> We are the first generation with widespread knowledge of how our activities influence the Earth System, and thus the first generation with the power and the responsibility to change our relationship with the planet. Responsible stewardship entails emulating nature in terms of resource use and waste transformation and recycling, and the transformation of agricultural, energy and transport systems.[60]

While we may query language which purports that humans are external to nature, or that awareness of global ecological issues is as novel as is claimed, the general notion of mobilizing widespread action to reconfigure agricultural, energy and transportational systems away from contemporaneous carbon-intensive models, and towards zero-waste alternatives is worthy of our solidarity. We must emphasize, however, that achieving such goals is not purely dependent on science and technocratic resource management, but requires a new collective political project which must provide a radical departure from the presently hegemonic ideology of neoliberal capitalism which fetishizes competitive individualism and efficiency over resilience, equity and commonwealth.

Such a project must be wary of the universalized metanarrative and teleology which currently permeates discourses of the Anthropocene, whereby small-scale interventions and local ecosystem alterations geared towards resilience (see Stevens, O'Brien, Holm this volume) disappear within the collective panic generated by the metanarrative of ecocide. This is not to suggest that individualized actions, such as taking shorter showers or using energy saving lightbulbs will avert the impacts of climate change, but that a diverse array of locally implemented actions are needed to build the communal and ecological resilience required to bind together multispecies assemblages in the troubled times which are forecast. While it may be too late to avoid significant changes to the climate or biodiversity loss, decisions taken over the coming years influence the extent of these changes and therefore the evolving conditions under which the politics of the Anthropocene will unfold.

NOTES

1. Paul Crutzen and Eugene Stoermer, "The 'Anthropocene.' Global Change Newsletter 41, 17–18," *International Geosphere–Biosphere Programme (IGBP)* (2000). Crutzen won the 1995 Nobel Prize in Chemistry for his work into the depletion of stratospheric ozone, and it is worth highlighting that the modern inception of the term traces to an individual whose scientific work directly led to concerted international action being taken to remedy human actions that posed a serious threat to the stability of planetary ecological systems.

2. Damian Carrington, "The Anthropocene Epoch: Scientists Declare the Damn of Human-Influenced Age," *The Guardian*, 29th August 2016, https://www.theguardian.com/environment/2016/aug/29/declare-anthropocene-epoch-experts-urge-geological-congress-human-impact-earth.

3. While the term "Anthropocene" can been various translated as the epoch or age of humans, mankind or man, I am using Elizabeth Kolbert's formulation of the Age of Man here. Kolbert, Elizabeth, "Enter the Anthropocene – Age of Man," *National Geographic* 219, no. 3 (2011): 60.

4. Gilles Deleuze and Félix Guattari, *A Thousand Plateaus: Capitalism and Schizophrenia* (Minneapolis: University of Minnesota Press, 1987), 2.

5. Donna Haraway, "Anthropocene, Capitalocene, Plantationocene, Chthulucene: Making Kin," *Environmental Humanities* 6 (2015): 160.

6. Bruce D. Smith and Melinda A. Zeder, "The Onset of the Anthropocene," *Anthropocene* 4 (2013): 6.

7. William F. Ruddiman, "The Anthropocene," *Annual Review of Earth and Planetary Sciences* 41, no. 1 (2013): 45–68; William F. Ruddiman, "The Early Anthropogenic Hypothesis: Challenges and Responses," *Reviews of Geophysics* 45, no. 4

(2007): 1–37; William F. Ruddiman, "The Anthropogenic Greenhouse Era Began Thousands of Years Ago," *Climatic change* 61, no. 3 (2003): 261–93.

8. William F. Ruddiman, Zhengtang Guo, Xin Zhou, Hanbin Wu, and Yanyan Yu, "Early Rice Farming and Anomalous Methane Trends," *Quaternary Science Reviews* 27, no. 13 (2008): 1291–95.

9. Ruddiman, "The Anthropogenic Greenhouse Era Began Thousands of Years Ago."

10. Martin Claussen, Victor Brovkin, Reinhard Calov, Andrey Ganopolski, and Claudia Kubatzki, "Did Humankind Prevent a Holocene Glaciation?" *Climatic Change* 69, no. 2–3 (2005): 409–17; James Hansen, Makiko Sato, Paul Hearty, Reto Ruedy, Maxwell Kelley, Valerie Masson-Delmotte, Gary Russell, et al., "Ice Melt, Sea Level Rise and Superstorms: Evidence from Paleoclimate Data, Climate Modeling, and Modern Observations That 2° C Global Warming Is Highly Dangerous," *Atmospheric Chemistry and Physics Discussions* 15, no. 14 (2015): 20059–179.

11. Jan Zalasiewicz, Colin N. Waters, Mark Williams, Anthony D. Barnosky, Alejandro Cearreta, Paul Crutzen, Erle Ellis, et al., "When Did the Anthropocene Begin? A Mid-Twentieth Century Boundary Level Is Stratigraphically Optimal," *Quaternary International* 383 (2015): 196–203, 198.

12. Andreas Malm, "The Anthropocene Myth," *Jacobin*, 2015, https://www.jacobinmag.com/2015/03/anthropocene-capitalism-climate-change/.

13. Clive Hamilton and Jacques Grinevald, "Was the Anthropocene Anticipated?" *The Anthropocene Review* 2 (2015): 62.

14. Simon L. Lewis and Mark A. Maslin, "Defining the Anthropocene," *Nature* 519, no. 7542 (2015): 171–80.

15. Nathan Nunn and Nancy Qian, "The Columbian Exchange: A History of Disease, Food, and Ideas," *The Journal of Economic Perspectives* 24, no. 2 (2010): 163–88.

16. Immanuel Wallerstein, *The Capitalist World-Economy*, Vol. 2. Cambridge: Cambridge University Press, 1979.

17. Jan Zalasiewicz, "Epochs: Disputed Start Dates for Anthropocene," *Nature* 520, no. 7548 (2015): 436.

18. Clive Hamilton, "Getting the Anthropocene So Wrong," *The Anthropocene Review* 2, no. 2 (2015): 102–07.

19. Crutzen and Stoermer, "The 'Anthropocene'."

20. Zalasiewicz, et al., "When Did the Anthropocene Begin?"

21. Will Steffen, Wendy Broadgate, Lisa Deutsch, Owen Gaffney, and Cornelia Ludwig, "The Trajectory of the Anthropocene: The Great Acceleration," *The Anthropocene Review* 2 (2015): 81–98.

22. Jan Zalasiewicz, Mark Williams, Colin N. Waters, Anthony D. Barnosky, and Peter Haff, "The Technofossil Record of Humans," *The Anthropocene Review* 1, no. 1 (2014): 34–43.

23. Zalasiewicz, et al., "When Did the Anthropocene Begin?"; Steffen, et al., "The Trajectory of the Anthropocene"; Colin N. Waters, Jan Zalasiewicz, Colin Summerhayes, Anthony D. Barnosky, Clément Poirier, Agnieszka Gałuszka, Alejandro Cearreta, et al., "The Anthropocene Is Functionally and Stratigraphically Distinct from the Holocene," *Science* 351, no. 6269 (2016): 137–49.

24. Michel Foucault, *The Foucault Reader*. New York: Pantheon, 1984, 81.

25. Rodney Harrison, "Surface Assemblages: Towards an Archaeology in and of the Present," *Archaeological Dialogues* 18, no. 2 (2011): 141–61; Paul Graves-Brown, Rodney Harrison, and Angela Piccini, *The Oxford Handbook of the Archaeology of the Contemporary World*. Oxford: Oxford University Press, 2013; Edward Cecil Harris, "Archaeological Stratigraphy: A Paradigm for the Anthropocene," *Journal of Contemporary Archaeology* 1, no. 1 (2014): 105–09.

26. Paul Graves-Brown, "When Was the Anthropocene? (and Why?)," *Journal of Contemporary Archaeology* 1, no. 1 (2014): 77–81.

27. Mike Walker, Sigfus Johnsen, Sune Olander Rasmussen, Trevor Popp, Jørgen-Peder Steffensen, Phil Gibbard, Wim Hoek, et al., "Formal Definition and Dating of the Gssp (Global Stratotype Section and Point) for the Base of the Holocene Using the Greenland Ngrip Ice Core, and Selected Auxiliary Records," *Journal of Quaternary Science* 24, no. 1 (2009): 3–17.

28. Bruno Latour and Steve Woolgar, *Laboratory Life: The Construction of Scientific Facts*. London: Sage, 1979; Bruno Latour, *Pandora's Hope: Essays on the Reality of Science Studies*. Cambridge, MA: Harvard University Press, 1999; John Law and Annemarie Mol, "Situating Technoscience: An Inquiry into Spatialities," *Environment and Planning D: Society and Space* 19, no. 5 (2001): 609–21.

29. Manuel Castells, *The Rise of the Network Society: The Information Age: Economy, Society and Culture Volume I*. Massachusetts: Blackwell, 1996; Jodi Dean, *Democracy and Other Neoliberal Fantasies: Communicative Capitalism and Left Politics*. Durham, NC: Duke University Press, 2009.

30. Bernard Stiegler, *What Makes Life Worth Living: On Pharmacology*. Cambridge: Polity, 2013.

31. Bernard Stiegler, "Automatic Society 1: The Future of Work, Introduction," *La Deleuziana – Online Journal of Philosophy* 1 (2015): 130.

32. David Beer, "Productive Measures: Culture and Measurement in the Context of Everyday Neoliberalism," *Big Data & Society* 2, no. 1 (2015): 1–12.

33. Nigel Gilbert and Rosaria Conte, *Artificial Societies: The Computer Simulation of Social Life*. London: Taylor & Francis, Inc., 1995, 2.

34. Rajendra K. Pachauri, M. R. Allen, V. R. Barros, J. Broome, W. Cramer, R. Christ, J. A. Church, et al., "Climate Change 2014: Synthesis Report. Contribution of Working Groups I, Ii and Iii to the Fifth Assessment Report of the Intergovernmental Panel on Climate Change," (2014), 10.

35. Robert M. DeConto and David Pollard, "Contribution of Antarctica to Past and Future Sea-Level Rise," *Nature* 531, no. 7596 (2016): 591–97.

36. Matthew O'Brien, "Everything You Need to Know about High Frequency Trading: Why the Algobots that Rule Wall Street are Good-and Why They're Evil Too," *The Atlantic*, 2014. http://www.theatlantic.com/business/archive/2014/04/everything-you-need-to-know-about-high-frequency-trading/360411/.

37. David Lyon, "Surveillance, Snowden, and Big Data: Capacities, Consequences, Critique," *Big Data & Society* 1, no. 2 (2014): 1–13; Jonathan Crary, *24/7: Late Capitalism and the Ends of Sleep*. London and New York: Verso, 2013.

38. Sean Cubitt, Robert Hassan, and Ingrid Volkmer, "Does Cloud Computing Have a Silver Lining?" *Media, Culture & Society* 33, no. 1 (2011): 149–58; Sy Taffel,

"Escaping Attention: Digital Media Hardware, Materiality and Ecological Cost," *Culture Machine* 13 (2012): 1–28.

39. Jussi Parikka, "A Geology of Media." Minneapolis: University of Minnesota Press, 2015.

40. Jason W. Moore, *Capitalism in the Web of Life: Ecology and the Accumulation of Capital*. New York: Verso Books, 2015.

41. Oxfam, "Working for the Few: Political Capture and Economic Inequality." Oxford: Oxfam International, 2014.

42. Kate Raworth, "Must the Anthropocene Be a Manthropocene," *The Guardian*, 2014.

43. Robert J. Nicholls and Anny Cazenave, "Sea-Level Rise and Its Impact on Coastal Zones," *Science* 328, no. 5985 (2010): 1517–20.

44. Agata Durkalec, Chris Furgal, Mark W. Skinner, and Tom Sheldon, "Climate Change Influences on Environment as a Determinant of Indigenous Health: Relationships to Place, Sea Ice, and Health in An inuit Community," *Social Science & Medicine* 136–137 (July 2015): 17–26.

45. Bobby Subhabrata Banerjee, "Whose Land Is It Anyway? National Interest, Indigenous Stakeholders, and Colonial Discourses: The Case of the Jabiluka Uranium Mine," *Organization & Environment* 13, no. 1 (2000): 3–38.

46. Will Steffen, Jacques Grinevald, Paul Crutzen, and John McNeill, "The Anthropocene: Conceptual and Historical Perspectives," *Philosophical Transactions of the Royal Society of London A: Mathematical, Physical and Engineering Sciences* 369, no. 1938 (2011): 842–67.

47. Bill McKibben, *The End of Nature*. New York: Anchor Books, 1989.

48. Donna Haraway, *A Manifesto for Cyborgs: Science, Technology, and Socialist Feminism in the 1980s*. Center for Social Research and Education, 1985; Donna Haraway, *The Companion Species Manifesto: Dogs, People, and Significant Otherness*. Chicago: Prickly Paradigm Press, 2003.

49. Haraway, "Anthropocene, Capitalocene, Plantationocene, Chthulucene," 160.

50. Haraway, "Anthropocene, Capitalocene, Plantationocene, Chthulucene," 159.

51. McKenzie Wark, *Molecular Red: Theory for the Anthropocene*. London: Verso Books, 2015.

52. Joanna Zylinska, *Minimal Ethics for the Anthropocene*. Ann Arbor: Open Humanities Press, 2014.

53. Wark, *Molecular Red*.

54. Naomi Oreskes and Erik M. Conway. *Merchants of Doubt: How a Handful of Scientists Obscured the Truth on Issues from Tobacco Smoke to Global Warming* (USA: Bloomsbury Publishing, 2011); Michael E. Mann, *The Hockey Stick and the Climate Wars: Dispatches from the Front Lines*. New York: Columbia University Press, 2013.

55. Jean-François Lyotard, *The Postmodern Condition: A Report on Knowledge*, Vol. 10. Minneapolis: University of Minnesota Press, 1984.

56. David Harvey, *A Brief History of Neoliberalism*. Oxford: Oxford University Press, 2005.

57. Franco Berardi, *Precarious Rhapsody*. London: Minor Compositions, 2009.

58. Paul Virilio, *The Information Bomb*. London and New York: Verso, 2005.
59. Gilles Deleuze, "Postscript on the Societies of Control," *October* 59 (1992): 3–7.
60. Steffen, et al., "The Anthropocene."

BIBLIOGRAPHY

Banerjee, Bobby Subhabrata. "Whose Land Is It Anyway? National Interest, Indigenous Stakeholders, and Colonial Discourses: The Case of the Jabiluka Uranium Mine." *Organization & Environment* 13, no. 1 (2000): 3–38.

Beer, David. "Productive Measures: Culture and Measurement in the Context of Everyday Neoliberalism." *Big Data & Society* 2, no. 1 (2015): 1–12.

Berardi, Franco. *Precarious Rhapsody*. London: Minor Compositions, 2009.

Carrington, Damian. "The Anthropocene Epoch: Scientists Declare the Damn of Human-Influenced Age." *The Guardian*, 29th August 2016, https://www.theguardian.com/environment/2016/aug/29/declare-anthropocene-epoch-experts-urge-geological-congress-human-impact-earth.

Castells, Manuel. *The Rise of the Network Society: The Information Age: Economy, Society and Culture Volume I*. Massachusetts: Blackwell, 1996.

Claussen, Martin, et al. "Did Humankind Prevent a Holocene Glaciation?" *Climatic Change* 69, no. 2–3 (2005): 409–17.

Crary, Jonathan. *24/7: Late Capitalism and the Ends of Sleep*. London and New York: Verso, 2013.

Crutzen, Paul and Eugene Stoermer. "The 'Anthropocene.' Global Change Newsletter 41, 17–18." *International Geosphere–Biosphere Programme (IGBP)* (2000).

Cubitt, Sean, Robert Hassan, and Ingrid Volkmer. "Does Cloud Computing Have a Silver Lining?" *Media, Culture & Society* 33, no. 1 (2011): 149–58.

Dean, Jodi. *Democracy and Other Neoliberal Fantasies: Communicative Capitalism and Left Politics*. Durham, NC: Duke University Press, 2009.

DeConto, Robert M. and David Pollard. "Contribution of Antarctica to Past and Future Sea-Level Rise." *Nature* 531, no. 7596 (2016): 591–97.

Deleuze, Gilles. "Postscript on the Societies of Control." *October* 59 (1992): 3–7.

Deleuze, Gilles and Felix Guattari. *A Thousand Plateaus: Capitalism and Schizophrenia*. Minneapolis: University of Minnesota Press, 1987.

Durkalec, Agata, Chris Furgal, Mark W. Skinner, and Tom Sheldon. "Climate Change Influences on Environment as a Determinant of Indigenous Health: Relationships to Place, Sea Ice, and Health in an Inuit Community." *Social Science & Medicine* 136–137 (July 2015): 17–26.

Foucault, Michel. *The Foucault Reader*. New York: Pantheon, 1984, 81.

Gilbert, Nigel and Rosaria Conte. *Artificial Societies: The Computer Simulation of Social Life*. London: Taylor & Francis, Inc., 1995.

Graves-Brown, Paul. "When Was the Anthropocene? (and Why?)." *Journal of Contemporary Archaeology* 1, no. 1 (2014): 77–81.

Graves-Brown, Paul, Rodney Harrison, and Angela Piccini. *The Oxford Handbook of the Archaeology of the Contemporary World*. Oxford: Oxford University Press, 2013.

Hamilton, Clive and Jacques Grinevald. "Was the Anthropocene Anticipated?" *The Anthropocene Review* 2 (2015): 62.

———. "Getting the Anthropocene So Wrong." *The Anthropocene Review* 2, no. 2 (2015): 102–07.

Hansen, James, et al. "Ice Melt, Sea Level Rise and Superstorms: Evidence from Paleoclimate Data, Climate Modeling, and Modern Observations that 2° C Global Warming Is Highly Dangerous." *Atmospheric Chemistry and Physics Discussions* 15, no. 14 (2015): 20059–179.

Haraway, Donna. *A Manifesto for Cyborgs: Science, Technology, and Socialist Feminism in the 1980s*. Center for Social Research and Education, 1985.

———. *The Companion Species Manifesto: Dogs, People, and Significant Otherness*. Chicago: Prickly Paradigm Press, 2003.

———. "Anthropocene, Capitalocene, Plantationocene, Chthulucene: Making Kin." *Environmental Humanities* 6 (2015): 159–65.

Harris, Edward Cecil. "Archaeological Stratigraphy: A Paradigm for the Anthropocene." *Journal of Contemporary Archaeology* 1, no. 1 (2014): 105–09.

Harrison, Rodney. "Surface Assemblages: Towards an Archaeology in and of the Present." *Archaeological Dialogues* 18, no. 2 (2011): 141–61.

Harvey, David. *A Brief History of Neoliberalism*. Oxford: Oxford University Press, 2005.

Kolbert, Elizabeth. "Enter the Anthropocene – Age of Man." *National Geographic* 219, no. 3 (2011): 60.

Latour, Bruno. *Pandora's Hope: Essays on the Reality of Science Studies*. Cambridge, MA: Harvard University Press, 1999.

Latour, Bruno and Steve Woolgar. *Laboratory Life: The Construction of Scientific Facts*. London: Sage, 1979.

Law, John and Annemarie Mol. "Situating Technoscience: An Inquiry into Spatialities." *Environment and Planning D: Society and Space* 19, no. 5 (2001): 609–21.

Lewis, Simon L. and Mark A. Maslin. "Defining the Anthropocene." *Nature* 519, no. 7542 (2015): 171–80.

Lyon, David. "Surveillance, Snowden, and Big Data: Capacities, Consequences, Critique." *Big Data & Society* 1, no. 2 (2014): 1–13.

Lyotard, Jean-François. *The Postmodern Condition: A Report on Knowledge*, Vol. 10. Minneapolis: University of Minnesota Press, 1984.

Malm, Andreas. "The Anthropocene Myth." *Jacobin*, 2015, https://www.jacobinmag.com/2015/03/anthropocene-capitalism-climate-change/.

Mann, Michael E. *The Hockey Stick and the Climate Wars: Dispatches from the Front Lines*. New York: Columbia University Press, 2013.

McKibben, Bill. *The End of Nature*. New York: Anchor Books, 1989.

Moore, Jason W. *Capitalism in the Web of Life: Ecology and the Accumulation of Capital*. New York: Verso Books, 2015.

Nicholls, Robert J. and Anny Cazenave. "Sea-Level Rise and Its Impact on Coastal Zones." *Science* 328, no. 5985 (2010): 1517–20.

Nunn, Nathan and Nancy Qian. "The Columbian Exchange: A History of Disease, Food, and Ideas." *The Journal of Economic Perspectives* 24, no. 2 (2010): 163–88.

O'Brien, Matthew. "Everything You Need to Know about High Frequency Trading: Why the Algobots that Rule Wall Street are Good-and Why They're Evil Too." *The Atlantic*, 2014, http://www.theatlantic.com/business/archive/2014/04/everything-you-need-to-know-about-high-frequency-trading/360411/.

Oreskes, Naomi and Erik M. Conway. *Merchants of Doubt: How a Handful of Scientists Obscured the Truth on Issues from Tobacco Smoke to Global Warming*. USA: Bloomsbury Publishing, 2011.

Oxfam. "Working for the Few: Political Capture and Economic Inequality." Oxford: Oxfam International, 2014.

Pachauri, Rajendra K., et al. "Climate Change 2014: Synthesis Report." Contribution of Working Groups I, Ii and Iii to the *Fifth Assessment Report of the Intergovernmental Panel on Climate Change* (2014).

Parikka, Jussi. "A Geology of Media." Minneapolis: University of Minnesota Press, 2015.

Raworth, Kate. "Must the Anthropocene Be a Manthropocene." *The Guardian*, 2014.

Ruddiman, William F. "The Anthropogenic Greenhouse Era Began Thousands of Years Ago." *Climatic change* 61, no. 3 (2003): 261–93.

———. "The Anthropocene." *Annual Review of Earth and Planetary Sciences* 41, no. 1 (2013): 45–68.

———. "The Early Anthropogenic Hypothesis: Challenges and Responses." *Reviews of Geophysics* 45, no. 4 (2007): 1–37.

Ruddiman, William F., Zhengtang Guo, Xin Zhou, Hanbin Wu, and Yanyan Yu. "Early Rice Farming and Anomalous Methane Trends." *Quaternary Science Reviews* 27, no. 13 (2008): 1291–95.

Smith, Bruce D. and Melinda A. Zeder. "The Onset of the Anthropocene." *Anthropocene* 4 (2013): 6.

Steffen, Will, Jacques Grinevald, Paul Crutzen, and John McNeill. "The Anthropocene: Conceptual and Historical Perspectives." *Philosophical Transactions of the Royal Society of London A: Mathematical, Physical and Engineering Sciences* 369, no. 1938 (2011): 842–67.

Steffen, Will, Wendy Broadgate, Lisa Deutsch, Owen Gaffney, and Cornelia Ludwig. "The Trajectory of the Anthropocene: The Great Acceleration." *The Anthropocene Review* 2 (2015): 81–98.

Stiegler, Bernard. *What Makes Life Worth Living: On Pharmacology*. Cambridge: Polity, 2013.

———. "Automatic Society 1: The Future of Work, Introduction." *La Deleuziana – Online Journal of Philosophy* 1 (2015).

Taffel, Sy. "Escaping Attention: Digital Media Hardware, Materiality and Ecological Cost." *Culture Machine* 13 (2012): 1–28.

Virilio, Paul. *The Information Bomb*. London and New York: Verso, 2005.

Walker, Mike, et al. "Formal Definition and Dating of the Gssp (Global Stratotype Section and Point) for the Base of the Holocene Using the Greenland Ngrip Ice Core, and Selected Auxiliary Records." *Journal of Quaternary Science* 24, no. 1 (2009): 3–17.

Wallerstein, Immanuel. *The Capitalist World-Economy*, Vol. 2. Cambridge: Cambridge University Press, 1979.

Wark, McKenzie. *Molecular Red: Theory for the Anthropocene*. London: Verso Books, 2015.

Waters, Colin N., et al. "The Anthropocene Is Functionally and Stratigraphically Distinct from the Holocene." *Science* 351, no. 6269 (2016): 137–49.

Zalasiewicz, Jan. "Epochs: Disputed Start Dates for Anthropocene." *Nature* 520, no. 7548 (2015): 436.

Zalasiewicz, Jan, et al. "When Did the Anthropocene Begin? A Mid-Twentieth Century Boundary Level Is Stratigraphically Optimal." *Quaternary International* 383 (2015): 196–203.

Zalasiewicz, Jan, Mark Williams, Colin N. Waters, Anthony D. Barnosky, and Peter Haff. "The Technofossil Record of Humans." *The Anthropocene Review* 1, no. 1 (2014): 34–43.

Zylinska, Joanna. *Minimal Ethics for the Anthropocene*. Ann Arbor: Open Humanities Press, 2014.

Index

aboriginal, 74, 78, 83, 91, 92, 100–103, 229
activism, xvi, xx, 37, 45, 112, 120, 143, 150, 163, 182, 220
Adams, Ansel, 149–50
Adivasi, 129–32
Adorno, Theodor, 24
aesthetics, xvi–xviii, xxi, 11–12, 17–29, 41, 93, 144–45, 147, 149–50, 152, 154, 156, 162, 171–73, 180
Africa, 10, 75, 83, 129, 130, 221
agency, xiv, xvii–xix, 17–18, 23–29, 59–60, 76, 95, 108, 148, 156, 169, 182, 189, 220, 229, 232
agriculture, 75–70, 83, 89, 92, 94, 97–98, 108, 112, 116, 129, 131–32, 145, 147, 150–52, 198–99, 201–8, 231.
  See also farming
Albrecht, Glenn, 155
alienation, 73, 93, 108–9, 112, 114–16, 151, 228
Amundsen, Roald, 167–68
Antarctica, xi, xxi, 17, 159–73
anthropocene, xiv–xviii, 15–16, 18, 22–23, 27–30, 73–74, 79, 85, 91, 94, 107, 113, 115, 125, 134, 143–45, 149, 153, 178–79, 186, 219–33;
  Anthropocene Working Group, 219, 222, 224, 228, 232;
  capitalocene, 227–29;
  definition of, x–xi, 219–21;
  evidence for, xi–xiii, xxi, 219–25;
  Great Acceleration, 223–24;
  and the Holocene, xi, 221–22
anthropology, x, 125–26, 134
Aotearoa New Zealand, xvi, xviii–xx, 19, 48, 53, 55, 66, 73, 75, 79, 82, 110, 114;
  Crown, 35–37, 42–43, 55–59, 61–62, 66
Appadurai, Arjun, 25
Artists in Antarctica program, xxi, 162–64, 168
Asia, 10, 83, 129, 221
atmosphere, 7, 76, 84, 108, 222
Auckland, 117, 161
Australia, xix–xxi, 22–23, 73–78, 81–83, 91–92, 94, 99–103, 143–49, 151–53, 155–56, 197–209
avant garde, 5, 179
te awa. *See* river

Bachelard, Gaston, 40
Baer, Ulrich, 146
Banks, Joseph, 198
Barad, Karen, ix–x, 160
Baxter, James K., 45
beavers, 12

Beeker, Chip, 10–11
Berry, Wendell, 91, 94, 97–98, 102
big data, 225–26
biodiversity, xiv, xxi, 61, 82, 108, 110, 144, 198, 209–10, 233
Bird-Rose, Deborah, 100–101
Birds, 60, 103, 133, 165, 167, 201, 203–6
Blackfish, 178, 185
Botany Bay, 197–98
Brown, Nigel, 168
Buell, Laurence, 179, 181, 183, 187
the bush, 77–78, 92, 94, 96, 115, 147–49, 197, 199, 201–2, 206, 208–9

Calder, Alex, 45
capitalism, xi, xv, xviii, 3, 6, 8–12, 75, 98, 103, 109, 120, 127, 144, 179, 208, 225–26, 229, 232;
  neoliberalism, 36, 46, 108, 109, 116, 226, 232
carbon, xviii, 4, 6, 10, 11, 12, 95–96, 108, 114, 181, 222, 232;
  carbon dioxide, 11, 222–23;
  carbon monoxide, 6
Carson, Rachel, 185
Carter, Paul, 22–23
chemicals, 4, 12
cities, xi, xv, 4, 46, 83, 92, 155, 206, 226
climate, 26, 75, 92, 96, 98, 111, 133, 154, 163, 173, 198, 200, 202
climate change, xi–xiv, xxi, 5, 10, 20, 108, 114, 119, 128–29, 150, 155–56, 159–60, 198, 223, 226–27, 229, 233
coal, xviii, 3–16, 223. *See also* carbon
colonialism, 22–23, 25, 63, 148–50, 152, 197, 223, 228;
  colonial, xix, 4, 22, 39, 64, 66, 73–74, 75, 77–78, 80–83, 85, 100, 127–28, 148, 154, 222;
  colonies, 4, 75, 77, 147, 198–203;
  colonisation, 36, 63, 73, 77, 79–84, 141, 143, 146, 160, 223;
  colonised, 9, 77, 102, 159;
  postcolonial, xix, 127, 181
Commanche, 81–82
commodity, 7–11, 130, 144, 147, 152, 178, 226
commons, 9–10, 12, 58
communication, 3–4, 7, 12–13, 127, 145, 166–68, 171, 210, 231
community, xx, 20, 25, 27, 42–43, 45, 47–48, 107–21, 141, 187
compost, 95–97
conformism, 18, 20–26
conservationism, xvii, xx, 28, 110, 150, 153, 181, 185–86, 189, 197, 201–9, 228
consilience, 145–46
consumption, xiii, xv, 5–6, 8, 93, 109, 119–20, 150;
  of food, 113, 116, 125, 130–32
corporations, xiv, 9–10, 41, 48, 93, 119, 131, 152, 227, 228
Cousteau, Jacques, 184
creativity, 110, 120
Crown. *See* Aotearoa New Zealand
Crutzen, Paul, x, 219, 223

Darwin, Charles, 189
Dawson, Martin, 36–39, 41–42, 46–47
DDT (dichlorodiphenyltrichloroethane), 206–7
debt, 9–13, 45
deep ecology, xvii, 28
degradation, environmental, xiv, 22, 42, 85, 92–94, 200, 209, 211
Deleuze, Gilles and Felix Guattari, 81–82, 220
desertification, 81–83
digital media, 3, 12
dinosaurs, xviii, 188
domestication, 92, 100–101, 132, 145, 147, 197, 221
drought, 10, 81–82, 85, 92, 94, 96, 148, 154, 165
drylands, 75–76, 82–83, 94

dust bowl, 81–84, 155
dye, 4–5, 12, 198

earth system, xiii, 222–23, 229, 232
Eco, Umberto, 127
ecology, ix–xvii, xix–xxi, 17–29, 35, 37, 41, 44, 81–83, 91, 93–96, 102–3, 107–8, 111, 113, 115–16, 118, 125, 129–30, 143–50, 156, 160, 163, 168–69, 171, 186, 207, 210, 219–20, 223–33
economics, 5–8, 18–19, 28, 41, 44, 46, 83, 92, 98, 116, 127, 145, 148–50, 152, 156, 197–211, 228, 230–31
ecosystem, x–xiii, xviii, 38, 79, 83, 94, 96, 98, 101, 131, 153, 159–60, 162, 169–70, 197, 221–22, 227, 229, 233
electricity, 3, 6, 40–41, 55, 119. *See also* hydro-electricity
enlightenment, xi, 21, 63, 200, 202, 231–32
entanglement, ix–xi, xiii–xxi, 17–22, 25–30, 35–37, 43, 56, 61, 108, 110–13, 120, 125–28, 132, 134–35, 143, 155–56, 159–60, 163, 169–70, 220, 228–29
environment, ix–x, xv–xvii, xxi, 7, 10, 13, 17–29, 35, 40–42, 44, 48, 58, 62–63, 75, 79, 81–83, 91, 93, 97, 100, 107, 109–12, 116, 119–20, 127, 130, 134, 145–50, 152, 154, 156, 159–73, 180, 184–85, 197, 207–8, 210–11, 229
environmentalism, 81, 94, 97, 100, 102, 109, 113, 114, 143–45, 148–56, 163, 177–80, 185, 187, 198, 206, 209–11, 229
Environment Network Manawatu, 110
epistemology, ix, 61, 65, 210
Erebus, Mount, 161–62, 167
ethics, xvi–xvii, xix, 17, 21, 27, 47, 73–74, 91, 108–9, 111, 118, 134, 144, 156, 200, 230–32
Europe, 5, 116
Externalities, 3, 12, 226
extinction, xii–xiv, 17, 179, 182, 185–86, 188, 202, 205, 220–21, 223, 225, 230, 232;
  mass extinction event, xiii, xiv, 220, 221, 225, 230, 232

farming, 46, 75–78, 81, 84–86, 92, 94, 96, 99–100, 102, 107, 110, 112, 114–17, 125, 128–32, 155, 183, 199, 207–8, 222, 228
Farrell, Michael, 77–78
film, xxi, 12, 84–85, 170, 175, 177, 182–89
finance, 3, 7–9
fish, 36, 96, 162, 177–89, 200;
  sharks, 183, 185, 186, 187–88
fishing, xii, 44, 179, 183, 187–89, 228
Flipper, 185, 187–88
Foucault, Michel, x–xi, 224
fracking, 5–6
Freud, Sigmund, 11, 148–51
fruit, 116–18, 201, 204

gardens, 19–22, 76, 92, 97, 103, 107–19, 179, 201–2
genealogy, 39, 59, 60, 177, 224. *See also* whakapapa
geological time scale, xi, 219, 221, 230
geology, x, 9, 12, 153, 163, 178–79, 186, 219–25, 227–32
Governor, Jimmy, 78
grass, 18–23, 26–27, 29, 76, 81–82, 86, 94, 96, 121, 130, 163, 207
Great Barrier Reef, 163
Great Dust Bowl, 81–84, 155
greenhouse gases, xi, 222–23, 226–28
Green Revolution, 131–33, 136
(GSST) Global Stratotype Section and Point Marker, 219–24

Hall, Bernedette, 164–65, 172
Haraway, Donna, 133–34, 221, 229–30
Harvey, David, 8–9
Head, Lesley and Pat Muir, 17, 26

Heidegger, Martin, 94–95
Henare, Amiria, 56, 60, 64
holocene, xi, 221–24
hunting, 93, 167, 179, 182, 187, 203, 205
hydro-electricity, xix, 40–42

ideology, xi, 148–49, 152, 205, 232
Illich, Ivan, 40, 177
India, xx, 5, 117–34, 228
indigenous, xviii, xx, 35–36, 44, 46–47, 56–57, 61, 64, 73–75, 79, 82, 84–86, 91, 94, 101–2, 130, 145, 148, 197, 203, 211, 228.
  See also aboriginal and māori
industrial revolution, xi, xv, 79, 223
Ingold, Tim, 99
interdisciplinarity, ix, xiii, xviii, 108, 141, 145–46, 162, 171
internet, 7–8, 12, 171–72
Interstellar, 84
Irrigation, 18–19, 23, 40, 46, 92, 129

Jameson, Fredric, 24

kaitiakitanga, 35, 40, 57–60, 61, 62, 63–65
Kelly, Ned, 77–79
Ko Aotearoa Tēnei, 56, 57–58, 61, 65, 66
Kohn, Eduardo, 126
Kolbert, Elizabeth, x, xii, 179, 186, 233n3

labour, 6, 41, 63–64, 91, 93–94, 97, 98, 109, 198, 199;
  slave, 77
landscape, xv, xx–xxi, 20, 25, 26, 44, 82, 85, 91, 92, 94, 98, 101, 127, 143–56, 159, 163, 165, 166, 168, 170, 197–98, 200–201, 204–11
Latour, Bruno, 28, 65, 235n28
lawns, 8, 17–30, 202;
  lawnmowers, 19–20, 26, 28;
  lawn-mowing, 17–20, 22, 24, 26, 27–29
Lazzarato, Maurizzio, 8
Levinas, Emmanuel, 134, 138n32
Levi-Strauss, Claude, 178
local, xi, xii, xiii, xiv, xv, xvi, xvii, xx, 17, 20, 21, 22, 23, 24, 25, 26, 27, 83, 96, 102, 107–9, 111, 112, 113, 115, 116–21, 131, 135, 155, 160, 170, 208, 211, 231, 233
Locke, John, 63–64
London, 3, 4, 40, 188, 201
Lovelock, James, 169
Luxemburg, Rosa, 9

MacDonald, Dwight, 22
Main, George, 82–83, 85, 86
mana, 35, 38, 39, 42, 46, 59–60
Manawatu, vii, xx, 110, 112, 113, 114, 116, 117
Manhire, Bill, 163–64, 166–69, 173
Māori, xvii, xix, 35–49, 55–66, 79–80, 134
markets, xviii, 4, 5, 7–8, 9, 10, 13, 46, 91, 108, 109, 112, 118, 126, 136n19, 209, 210, 227;
Marshall, Owen, 165–66;
Marx, Karl, 8, 9, 179
mass extinction event. *See* extinction
Massey University, vii, xvi, 11, 114
materiality, x, 3, 7, 13, 17, 23, 26, 57, 60, 61, 74, 134, 137, 146, 156, 180, 186, 189, 225, 227, 229, 232
mateship, 148–49, 153
Mauss, Marcel, 60
McKibben, Bill, 229
media, 3, 4, 6, 7, 12, 13, 129, 178, 179, 185, 227, 230
mediation, 4, 64, 91, 96, 135, 152–53, 186, 189
Melville, Herman, 177–79, 181–83, 187
microbiota, xiv, xviii, 91, 95, 96, 125, 133–35, 170, 204, 230;
  bacteria, 97, 133, 230;
  viruses, 133
millet, 128–34, 136n18
mining, 3–6, 12, 93, 229
Mitchell, Timothy, 6, 181
Moby-Dick. *See* Melville, Herman

modernity, 5, 9, 12, 40, 56, 63, 64, 91–93, 98, 99, 102, 127, 145, 150, 178, 179, 181, 182, 187, 197, 206, 207–8, 223, 228, 231;
postmodern, 231–32;
pre-modern, 127
monoculture, xx, 22, 91, 98, 108, 131, 132
Moore, Jason, 228–29
mutualism, xv, xx, xxi, 99, 108–9, 127, 128, 143–44, 152, 156, 229

narrative, xvii, xix, xx, xxi, 11, 22, 65, 83, 125, 128, 132, 135, 143, 160–66, 168–72, 188, 219, 221, 230;
metanarrative, 231–33
national parks, 149, 150, 202, 208
nature, vii, ix, x, xii, xiii, xiv, xv, xvi–xviii, xix, xx, 5, 7, 11, 17, 18, 21, 22, 23, 25, 26, 27, 28, 29, 40, 43, 45, 48, 56, 63–65, 66, 73–75, 77, 78–79, 82–86, 108, 134, 145, 147, 148, 149–54, 159, 160–63, 169, 170, 171–72, 177, 178, 179, 180, 183, 184, 197, 200, 201, 202, 203, 205, 206, 207, 208, 210, 225, 228, 229, 230, 232;
nature/culture binary, ix, x, xix, 56, 63–64, 66, 143, 144–45, 155, 156, 229, 231
neoliberalism. *See* capitalism
networks, ix, 12, 13, 110, 113, 114, 117, 119, 129, 227
New South Wales, xxi, 74, 76, 83, 147, 153, 155, 199, 200, 201, 203, 204, 205, 207, 208
New Zealand. *See* Aotearoa New Zealand
nonhuman, ix–xi, xiv, xv, xvi, xvii, xviii, xix, xx, xxi, 13, 17, 18, 20, 21–23, 25–29, 39, 56–57, 64, 65–66, 110, 134, 143, 145, 156, 225, 228
nuclear energy, 5
nuclear weapons, xii, 223, 228

ocean. *See* sea
oil, 3, 5, 6, 11, 181, 182
ontology, x, xix, 56, 58, 61, 63, 64 ,65, 133, 134, 225, 229, 230, 231
Orsman, Chris, 166, 168, 169–70

Painleve, Jean, 183–84, 187
pākehā, 35, 36, 39, 42, 45, 46, 56, 57, 63, 64, 65, 66, 79–80
Palmerston North, 113, 119. *See also* Manawatu
penguins, 162, 165, 168, 170, 172
Perkin, William, 4
pests, xv, 75, 96, 108, 203, 204, 209, 210
photography, xx–xxi, 11, 143–53, 159, 169, 171
plants, xiii, xviii, 23, 26, 71, 79, 83, 92, 94, 95, 87, 101, 107, 111, 113, 116, 117, 119, 125, 132, 133, 145, 199, 202, 205, 209
Plant-to-Plate, 114–15, 118–19
plastic, xi, 160, 224, 227, 228
plough, xix, 73–86, 92, 96, 107
poetry, xxi, 45, 73, 77, 137n27, 161–73, 182, 202
Pollan, Michael, 21–22, 24–25
pollution, xiv, xvii, 1, 44, 46, 48, 178
posthumanism, x, 229, 232
privatisation, xix, 43, 55, 58, 67n1
Probyn, Elspeth, 187–88
property, 62, 63, 64, 66, 200;
intellectual, 36;
rights, xviii, xix, 43, 56, 57–58, 61, 63, 64, 65
psychology, xxi, 41, 143–53, 155

Ramsden, Irihapeti, 47–48
rationality, 62–63, 66, 127
reciprocity, xix, 60, 91, 96, 99, 100, 103
resilience, xiv, xx, 92, 94, 117, 125, 128, 132, 133, 162, 198, 222, 233
rice, 129, 130, 131, 132, 136n16, 222
Riley, Robert, 21
Ritchie, James, 41
rivers, xi, xv, xviii, xix, 79, 92. *See also* Whanganui River and Waikato River

Robbins, Paul and Julie Sharp, 18, 20
Ruru, Jacinta, 66

Science, x, xi, xiii, xiv, xvi, xxi, 64, 65, 66, 79, 83, 84, 102, 134, 155, 159–60, 162–64, 167–68, 171, 172, 173, 178, 183–84, 186, 206, 209, 210, 211, 219, 220, 221, 222, 224, 225, 226, 227, 229, 230, 231, 232
Scott, Robert, 161, 169;
  Scott base, 162, 163, 164, 167, 169, 170, 172
Sea, xi, xviii, xxi, 17, 35, 38, 43, 102, 160, 163, 177–89, 223, 230
Serres, Michel, 73, 75, 78–79
Sewell, Bill, 161
Sharks. *See* fish
Sheep, 94, 96, 100, 199
Shepard, Paul, 145, 151–52
simulation, xiii, xxi, 220, 226, 227
soil, xv, xx, 23, 74–75, 76, 78, 81, 82, 91–92, 94–97, 99, 100, 107, 108, 111, 114, 116, 117, 120, 130, 133, 199–200, 201, 207, 208–9
solastalgia, 155–56
Sontag, Susan, 154
Stiegler, Bernard, 225
Strang, Veronica, 41
sublime, 147, 148, 149, 156
suburbs, 17–22, 26
sustainability, 42, 117, 119, 128, 189, 204, 209–11
Sydney, 83, 92, 199, 201
symbiosis, 107–8, 125, 133, 134, 144, 153, 154, 229

taonga, 37, 42, 44, 56–66
Tapsell, Paul, 59–60
Taranaki, 79–80
Taussig, Michael, 188
Tawhiri, Taitoko, 41
Te Awa Tupua Agreement, 42–44, 45–46, 48
terraforming, 18, 201, 203, 230
Tongariro, 37, 41
trauma, 11, 146, 150–51, 154
Tuhiwai Smith, Linda, 62
Turia, Tariana, 39, 42

United States of America, xix, 5, 8, 20, 22, 28, 29, 73, 74, 75–76, 78, 81, 82, 83, 86, 116, 149, 151, 198, 222
use, xix–xx, 19, 27–28, 43, 58, 91–102;
  use value, 12;
  kindly use, 91, 94, 97–98, 102

Waikato River, 41, 55
Waitangi, Treaty of (*Tiriti ō Waitangi)*, xix, 35, 36, 37, 47, 48, 55, 59, 61;
  tribunal, 37, 38, 41, 42, 43, 55, 56, 57, 59, 61, 62, 64, 66, 67n1;
  wai, 262, 56, 57–61, 62–64, 65
war, 5, 10;
  American Civil War, 82;
  New Zealand Wars, 36;
  World War II, xii, 151, 180
Wark, McKenzie, 230
waste, xiv, 7, 117, 222, 232
water, xix, 4, 7, 23, 35–46, 48, 55–58, 61, 66, 75, 81, 92, 96, 108, 110, 126–27, 129, 136n16, 148, 155, 167, 177–78, 208
Watt, James, 223
The West, xix, 61, 63, 64, 65, 66, 73, 74, 77, 85, 99, 100, 102, 103, 143, 210, 228
whakapapa, 39, 59, 65
whales, 177, 178, 179, 180, 181–83, 185, 186, 188, 189, 199
Whanganui River, xix, 35–48
Whitehead, Alfred North, 126
Wilderness, xviii, xix, xx, 22, 27, 100, 126, 127, 147, 148, 149, 150, 153, 168, 171, 172, 173
Wilson, E. O., 189, 210
Winiata, Whatarangi, 47, 49
Wood, Keith, 40, 42, 46, 49
Worster, Donald, 40

YoHa, 3
Young, David, 48, 49

About the Contributors

Octavia Cade holds a PhD in science communication from the University of Otago, specializing in creative non-fiction writing, and is currently a Masters of Science student at the University of Waikato investigating seagrass restoration. She is a published poet and author of science fiction, and has recently published in the journal *ISLE*.

Sean Cubitt is professor of Film and Television at Goldsmiths, University of London, and Honorary Professorial Fellow at the University of Melbourne. His most recent books are *The Practice of Light: A Genealogy of Visual Media* (MIT 2014) and *Finite Media: Environmental Implications of Digital Technologies* (Duke 2016). He is the series editor for Leonardo Books at MIT Press.

Charles Dawson is the director of an interdisciplinary New Zealand programme for the US social justice/sustainability network HECUA and Founding Vice President (New Zealand) of the Association for the Study of Literature Environment and Culture (Australia and New Zealand) (www.aslec-anz.asn.au). He completed his PhD ("Writing the Memory of Rivers") at UBC, Canada, and is a graduate of the famed Whanganui Tira Hoe Waka. His articles and reviews on water issues feature in Australasian and North American journals and the recent book *Ecocriticism of the Global South* (eds. Slovic, Rangarajan and Sarveswaran, 2015).

Victoria Grieves is an ARC Indigenous Research Fellow at the University of Sydney, and is of Warraimay and Tasmanian descent. She is currently completing an ARC-funded project exploring the history of the Aboriginal family over space and time – and also beginning a project on the Children born of war in Australia during the war in the Pacific 1941–1944. Her research interests involve gender and race in settler colonial societies, the development of Indigenous knowledges in the academy, well-being and cultural heritage management.

James Holcombe has a degree in visual arts, majoring in photography, from Sydney College of the Arts (Sydney University), and has completed postgraduate studies in art/photography at the University of Newcastle. He has recently completed a PhD in fine art photography at Charles Sturt University. In 1996, Jamie was appointed Honorary Visiting Academic to the Centre for Electronic Art at the University of Middlesex in London, returning to CSU as co-founder of the then multimedia and photo-imaging discipline. He subsequently developed and implemented the BA (Photography), introduced on the Wagga Wagga campus in 2006.

Nicholas Holm is a lecturer in media studies at Massey University. His research is broadly concerned with the role of aesthetics – and in particular the aesthetics of popular culture – in shaping the political and social conditions of contemporary thought. His interest in ecocriticism is motivated by the ethical and aesthetic status of urban nature, which he has previously explored in articles addressing squirrels and possums. He is currently completing two books entitled *Advertising and Society: A Critical Introduction* and *Humour as Politics*.

Anne O'Brien is a PhD candidate in political and social thought at Australian Catholic University, North Sydney. She is interested in transforming the interface between economy and ecology, so that care for the land – and particularly the soil – is built into everyday practices of land stewardship. With a background in geography and philosophy, as well as community-based activism, she uses both empirical and philosophical analyses to draw out normative principles that structure "background understandings" sustaining the status quo and repertoires that transcend it.

Christopher Orchard has a degree in visual arts, majoring in new media, and completed honours degree and a Master of Arts by research from Charles Sturt University. Christopher is completing a PhD while concurrently studying a GradCert in Research through the University of Tasmania. Christopher has also completed two years of Bachelor of Psychological Science from the University of New England and is due to complete in eighteen months. Christopher is an internationally exhibiting photographer with work in several major institutional collections including the National Gallery of Australia.

David Orchard is a PhD candidate at Charles Sturt University investigating water relations in plant-fungus symbioses. He previously completed a Bachelor of Arts with first-class honours in English from the Australian National University, receiving a University Medal, and a Graduate Diploma in science from the University of New England. His research has focused on the adaptive capacity of native plants and their roles in weed and insect control.

Peter Orchard is an adjunct senior lecturer and visiting scientist at the Graham Centre for Agricultural Innovation, an alliance between Charles Sturt University and NSW Primary Industries. Prior to this he was responsible for the statewide management of research and extension in the NSW DPI Pasture and Rangelands section. Current research interests include the adaptation of native shrub and herbaceous legumes, enhancing understorey components of revegetation initiatives to develop and document the benefits of more complex landscape systems, and creating change through transdisciplinary engagement.

Jacob Otter is a PhD candidate in Cultural Studies at the University of Melbourne, Australia. He explores themes of settler colonialism, political ontology, and indigeneity in controversies from Aotearoa New Zealand and Australia. He is an adjunct at the School of Social and Political Sciences at Auckland University of Technology and has previously published on neoliberalism and colonisation in Aotearoa New Zealand.

Gareth Stanton is a media academic at Goldsmiths, University of London. His research interests lie in the areas of postcolonial theory, anthropology of the media and the interface between cultural studies and anthropology. He has recently been researching the relationship between media and the environment with a special focus on Bangladesh. He is currently writing a book to be called, "Media, Culture and the Environmental Turn."

Sharon Stevens works for two environmental NGOs, the Environment Network Manawatu (a charitable umbrella organization) and RECAP (which provides adult community education in sustainability and permaculture). Her publications include *A Place for Dialogue: Language, Land Use, and Politics in Southern Arizona* (2007) and, as co-editor, *Active Voices: Composing a Rhetoric of Social Movements* (2009), plus articles addressing environmental communication and social change.

Sy Taffel is a lecturer in Media Studies at Massey University, Aotearoa New Zealand. His research interests include political ecologies of digital media, digital media and political activism, the material impacts of media hardware, pervasive/locative media, software studies, and peer-to-peer production. He has published work in journals including *Convergence*, *Culture Machine*, and the *NECSUS European Journal of Media Studies*. He has also worked as a filmmaker and photographer with an interest in audiovisual representations of ecological and social justice movements.

Sita Venkateswar is a socio-cultural anthropologist in the School of People, Environment and Planning, Massey University. She incorporates critical, scholar-activist research methodologies, designated as Public Anthropology, informed by feminist and postcolonial theories. Her current research explores multispecies approaches to food resilience that focuses on millet cultivation in India.